BIOTECHNOLOGICAL APPLICATIONS OF JASMONATES

A Group of Less Well-Known Plant Growth Regulators

BIOTECHNOLOGICAL APPLICATIONS OF JASMONATES

A Group of Less Well-Known Plant Growth Regulators

Tariq Ahmad Dar, PhD
Bilal Ahmad, PhD
Khalid Rehman Hakeem, PhD
Editors

APPLE ACADEMIC PRESS

First edition published 2025

Apple Academic Press Inc.
1265 Goldenrod Circle, NE,
Palm Bay, FL 32905 USA

760 Laurentian Drive, Unit 19,
Burlington, ON L7N 0A4, Canada

CRC Press
2385 NW Executive Center Drive,
Suite 320, Boca Raton FL 33431

4 Park Square, Milton Park,
Abingdon, Oxon, OX14 4RN UK

Library and Archives Canada Cataloguing in Publication

Title: Biotechnological applications of jasmonates : a group of less well-known plant growth regulators / Tariq Ahmad Dar, PhD, Bilal Ahmad, PhD, Khalid Rehman Hakeem, PhD, editors.
Names: Dar, Tariq Ahmad, editor. | Ahmad, Bilal (Lecturer at the Government Degree College for Women), editor. | Hakeem, Khalid Rehman, editor
Description: First edition. | Includes bibliographical references and index.
Identifiers: Canadiana (print) 20250133164 | Canadiana (ebook) 20250133199 | ISBN 9781774919569 (hardcover) | ISBN 9781774919576 (softcover) | ISBN 9781003595694 (ebook)
Subjects: LCSH: Jasmonic acid. | LCSH: Plant regulators. | LCSH: Plants—Development.
Classification: LCC QK745 .B56 2025 | DDC 571.7/2—dc23

Library of Congress Cataloging-in-Publication Data

CIP data on file with US Library of Congress

ISBN: 978-1-77491-956-9 (hbk)
ISBN: 978-1-77491-957-6 (pbk)
ISBN: 978-1-00359-569-4 (ebk)

Dedication

This book is dedicated to
my beloved Mother "Hajra Begum."
May Allah bless her Soul (Aameen)

About the Editors

Tariq Ahmad Dar, PhD
Assistant Professor Botany, Sri Pratap College; Srinagar (affiliated with Cluster University Srinagar), India

Dr. Tariq Ahmad Dar earned his MSc in Botany from the P.G. Department of Botany, University of Kashmir, Hazratbal, Srinagar, in 2010 and PhD in Botany from Aligarh Muslim University, Aligarh, India, in 2015. Dr. Dar has 20 publications to his credit. He has published his research findings in journals of international repute, including *Environmental and Experimental Botany, Industrial Crops and Products, Carbohydrate Polymers, Journal of Applied Research on Medicinal and Aromatic plants, Journal of Radiation Research and Applied Sciences, Plant Physiology Reports, Journal of Essential Oil Research, Frontiers in Plant Science, Journal of Plant Growth Regulation* etc. Dr. Dar has presented his papers at the national and international conferences and has attended about 11 national and international conferences. Moreover he is a life member of the Indian Botanical Society and Indian Science Congress Association. He has guided the dissertations of 10 MSc students. He is actively involved on the aspect of enhancing growth, yield, and active constituents of medicinal and aromatic plants using plant growth regulators like jasmonates, gibberellins, and depolymerized polysaccharides. Currently, Dr. Dar is focusing on the use of marine polysaccharides as plant growth regulators and increasing the yield of active constituents of medicinal plants. Also the use of radiation processed depolymerized polysaccharides like sodium alginate, chitosan, carageenan to alleviate heavy metal stress and different abiotic stress conditions like salinity, drought, chilling, etc. are focus areas. Dr. Dar has teaching experience of eight years at under graduate level and 2.5 years at postgraduate level.

Bilal Ahmad, PhD
Assistant Professor Botany, Higher Education Department of Jammu and Kashmir, India

Bilal Ahmad, PhD, is currently working as an Assistant Professor, Higher Education Department of Jammu and Kashmir, India. After completing his masters in Botany from Aligarh Muslim University, Aligarh, Uttar Pradesh, India, Dr. Ahmad pursued his PhD in Botany in the field of Advanced Plant Physiology till 2018. Dr. Ahmad has studied the influence of various plant growth promoters, radiation processed polysaccharides, and nanoparticles on various medicinal plants under normal and perturbed soil environments particularly heavy metal stress. Dr. Ahmad has been associated with teaching various disciplines of botany, notably plant physiology and biochemistry, cell and molecular biology, and plant ecology since 2018. Dr. Ahmad has presented his research at a number of national and international conferences and has received several awards. Dr. Ahmad has authored 36 research publications in high impact factor journals and edited volumes of different books with international publishers. In addition, Dr. Ahmad has also reviewed various articles for high impact factor journals of international repute.

Khalid Rehman Hakeem, PhD
Professor, King Abdulaziz University, Jeddah, Saudi Arabia

Khalid Rehman Hakeem, PhD, is working as a Professor at King Abdulaziz University, Jeddah, Saudi Arabia. He is currently the Head of the Plant Biotechnology Research Unit at the Princess Dr. Najlaa Bint Saud Al Saud Center for Distinguished Research in Biotech, King Abdulaziz University, Jeddah. Currently, he is also working as a Visiting Professor at Daffodil International University, Dhaka, Bangladesh. After completing his PhD in Botany, specializing in Plant Ecophysiology and Molecular Biology, from Jamia Hamdard, New Delhi, India, in 2011, he initially worked as an Assistant Professor at the University of Kashmir, Srinagar. Later, he joined Universiti Putra Malaysia, Selangor, Malaysia, and worked there as a Post-doctorate Fellow in 2012 and as a Fellow Researcher (Associate Professor) from 2013 to 2016, respectively. Dr. Hakeem joined King Abdulaziz

University in August 2016 as an Associate Professor and was promoted to Professor in 2019. Prof. Hakeem has more than 14 years of teaching and research experience in plant eco-physiology, biotechnology and molecular biology, medicinal plant research, plant-microbe-soil interactions, as well as environmental studies. He is the recipient of several fellowships at both national and international levels. He was recently elected as a Fellow of the Royal Society of Biology, London. He has served as a visiting scientist at Fatih University, Istanbul, Turkey, as well as at Jinan University, Guangzhou, China. Currently, he is involved with several international research projects with different government organizations. To date, he has authored and edited more than 100 books with international publishers, including Springer Nature, Academic Press (Elsevier), CRC Press, Apple Academic Press, etc. He has also to his credit more than 200 research publications in peer-reviewed international journals and over 90 book chapters in edited volumes with international publishers. At present, he is serving as an editorial board member and reviewer for several high-impact international scientific journals from Elsevier, Springer Nature, Taylor & Francis, Cambridge Press, and John Wiley. He is included in the advisory board of Cambridge Scholars Publishing, UK. He is also a Fellow of the Plantae Group of the American Society of Plant Biologists, a member of the World Academy of Science, a member of the International Society for Development and Sustainability, Japan, and a member of the Asian Federation of Biotechnology, Korea. He has been listed in *Marquis Who's Who in the World* from 2014 to 2021. Currently, he is engaged in studying plant processes at eco-physiological as well as molecular levels.

Contents

Contributors

Md. Adil
Department of Biotechnology, University of Kashmir, Srinagar, Jammu and Kashmir, India

Bilal Ahmad
Advanced Plant Physiology Section, Department of Botany, Aligarh Muslim University, Aligarh, Uttar Pradesh, India; Assistant Professor Botany, Higher Education Department Jammu and Kashmir

Asif Amin
Department of Biotechnology, University of Kashmir, Srinagar, Jammu and Kashmir, India

Nahila Anjum
School of Biosciences and Biotechnology, Baba Ghulam Shah Badshah University, Rajouri, Jammu and Kashmir, India

Kaisar Ahmad Bhat
School of Biosciences and Biotechnology, Baba Ghulam Shah Badshah University, Rajouri, Jammu and Kashmir, India; Proteomics Laboratory, Division of Plant Biotechnology, Sher-e-Kashmir University of Agricultural Sciences and Technology of Kashmir (SKUAST-K), Shalimar, Jammu and Kashmir, India

Ishani Chakrabartty
Learning and Development Solutions, Medical Affairs, Indegene Pvt. Ltd., Manyata Tech Park, Nagarwara, Bangalore, Karnataka, India

Tariq Ahmad Dar
Assistant Professor Botany, Sri Pratap College; Srinagar (affiliated with Cluster University Srinagar), India

Faizah Farooq
Department of Biotechnology, University of Kashmir, Srinagar, Jammu and Kashmir, India

Iqra Farooq
CSIR–Indian Institute of Integrative Medicine, Field Station, Pulwama, Jammu and Kashmir, India

Sajad A. Ganie
Division of Entomology, Sher-e-Kashmir University of Agricultural Sciences, Shalimar, Jammu and Kashmir, India

Yashodhara Goswami
Department of Biomedical Genetics, School of Biosciences and Technology, Institute of Technology, Vellore Campus, Vellore, Tamil Nadu, India

K. R. Hakeem
Department of Biological Sciences, Faculty of Science, King Abdulaziz University, Jeddah, Saudi Arabia

Mohd. Hanief
School of Biosciences and Biotechnology, Baba Ghulam Shah Badshah University, Rajouri, Jammu and Kashmir, India

Syed Humaira
Department of Botany, University of Kashmir, Jammu and Kashmir, India

Khursheed Hussain
Department of Botany, University of Kashmir Hazratbal, Srinagar, Jammu and Kashmir, India

Ajmat Jahan
Advanced Plant Physiology Section, Department of Botany, Aligarh Muslim University, Aligarh, Uttar Pradesh, India

Arjumand John
Department of Zoology, Lovely Professional University, Phagwara, Jalandhar, Punjab, India

A. Pravin Kumar
CSIR–Indian Institute of Integrative Medicine, Field Station, Pulwama, Jammu and Kashmir, India

Aashiq Hussain Mir
Department of Zoology, Lovely Professional University, Phagwara, Jalandhar, Punjab, India

Urwat Ul Muskan
Department of Botany, University of Kashmir, Jammu and Kashmir, India

Raies A. Qadri
Department of Biotechnology, University of Kashmir, Srinagar, Jammu and Kashmir, India

Vaseem Raja
University Center for Research and Development, Chandigarh University, Mohali, Punjab, India

Unjman Rashid
Department of Botany, Jamia Hamdard University, New Delhi, India

Shahid Rasool
CSIR–Indian Institute of Integrative Medicine, Field Station, Pulwama, Jammu and Kashmir, India

Aryadeep Roychoudhury
Discipline of Life Sciences, School of Sciences, Indira Gandhi National Open University, Maidan Garhi, New Delhi, India

Santanu Samanta
Post-Graduate Department of Biotechnology, St. Xavier's College (Autonomous), Mother Teresa Sarani, Kolkata, West Bengal, India

Ali Asghar Shah
School of Biosciences and Biotechnology, Baba Ghulam Shah Badshah University, Rajouri, Jammu and Kashmir, India

Umar Majeed Wani
Department of Biotechnology, University of Kashmir, Srinagar, Jammu and Kashmir, India

Sajad Majeed Zargar
Proteomics Laboratory, Division of Plant Biotechnology, Sher-e-Kashmir University of Agricultural Sciences and Technology of Kashmir (SKUAST-K), Shalimar, Jammu and Kashmir, India

Preface

Plant physiology and metabolism are controlled by a vast array of intrinsic signals of diverse nature, which allows them to flourish under normal and perturbed environmental conditions. Among these, phytohormones constitute an imperative class of signaling molecules that integrate endogenous developmental cues with environmental signals to regulate plant growth, development, and defense.

Jasmonates constitute a well-acknowledged class of lipid-derived phyto-regulators that enhance the adaptability of plants to biotic (herbivore attack, pathogen infection) as well as abiotic (wounding, heavy metals, salinity, ozone, temperature, ultraviolet radiation) stresses, in addition to regulating key growth and developmental phenomena. Like those of prostaglandins in mammals, JAs are cyclopentanone derivatives derived from α-linolenic acid through the pathway named the octadecanoid pathway.

The α-linolenic acid produced in the plastids is converted into jasmonic acid through a sequence of reactions, which is subsequently exported to the cytoplasm, where it can be activated or inactivated depending upon the physiological state of the cell, through a sequence of reactions like methylation, glucosylation, or sulfation. The endogenous concentrations of jasmonates in plants vary as a function of cell type and developmental stage, as well as in response to varied environmental stimuli.

Elevated concentrations of jasmonates have been observed in flowers, fruit pericarp, and in the plastids, particularly chloroplasts of illuminated plants, as well as in plant tissues exposed to mechanical perturbations like tendril coiling and wounding. The functional role of JAs in plant developmental phenomena, notably primary root growth, leaf senescence, and reproductive development, is well established.

Among the growth and developmental phenomena, the notable effect of JAs is during the regulation of lateral and adventitious root development. Whereas lateral root formation in Arabidopsis is promoted, adventitious root formation is inhibited. JAs attenuate the inhibitory effect on root growth by recruiting a co-repressor, which in turn inhibits the downstream transcription factors. Lateral root formation by JAs involves crosstalk with endogenous signaling molecules that up-regulate the promoters of auxin biosynthetic genes.

Research studies during the past two decades have focused on their functional role in diverse developmental phenomena, fluctuations in their biosynthesis under normal and perturbed environments, and their interplay with miscellaneous signaling molecules, including diverse phytohormones. These studies are the need of the hour in view of their potential to manipulate the genetic makeup of various other plants to make them better adapted to constantly changing environments.

Jasmonates inhibit the germination of non-dormant seeds but stimulate the germination of dormant seeds, encourage the accumulation of storage proteins, and up-regulate the expression of vegetative storage proteins and tuber formation in potatoes via a jasmonate derivative, tuberonic acid. Exogenous application of jasmonates down-regulates the expression of photosynthetic proteins, which aids the plant in trimming down carbon assimilation during conditions of light and other abiotic stresses.

The accumulation of jasmonates in flowers as well as fruits remains poorly understood; it is believed to be associated with the ripening of fruit (via ethylene), fruit carotenoid composition, and the expression of genes encoding vegetative and seed-storage proteins. Jasmonates increase the resistance of plants against stress by up-regulating the defense-related genes, and the synergism of JAs with ethylene during such defense is well established.

The signaling pathways of different phytohormones are latticed in a complicated network, imparting the plants with massive regulatory power to swiftly adapt to the variable environment, facilitating them to efficiently exploit the limited resources, which otherwise can negatively affect growth, development, and adaptation. JAs do not act discretely but rather in a complex relationship with other phytohormones, involving signaling crosstalk with them. Methyl jasmonate, a widely distributed jasmonate, is intangible with regard to its biogenesis, regulation, signaling cross-talk, and mode of action.

One of the imperative breakthroughs that helped in gaining insights into the role of MeJA in gene up-regulation as well as long-distance systemic signaling evolved with the characterization of a vital gene encoding jasmonic acid carboxyl methyl-transferase. Functional genomics and bioinformatic studies of JAs have helped in the identification of a set of MeJA-responsive genes, as well as the use of diverse volatile signals by plants, which help them endure miscellaneous and unpredictable environments.

Comprehensive studies have revealed intensive crosstalk (synergistic and/or antagonistic) between the signal transduction pathways of various other phytohormones and JAs to achieve the function of stress tolerance, both biotic and abiotic. However, many voids currently exist in comprehending jasmonate cross-talk, particularly in the perception of multiple stress signals. Nevertheless, omics studies are currently being focused on elucidating the composite protein interface to identify the multifaceted stress signaling pathways and diverse signaling proteins, messengers, and TFs-mediated jasmonate signaling cross-talk.

The book highlights the key advancements in jasmonate research in the past two decades, including the role of jasmonates in the regulation of growth and development, biosynthetic pathways, the role of jasmonates during stressful conditions, and the signal transduction cascades involved during the jasmonate-induced regulation of growth and development. The book will prove to be an ideal information source for a wide scientific audience, including students and research scholars.

Lastly, we extend our heartfelt appreciation to AAP (CRC Press, Taylor & Francis Group) for their extraordinary support in helping us achieve this success.

—Tariq Ahmad Dar, PhD
Bilal Ahmad, PhD
Khalid Rehman Hakeem, PhD

Introduction

Plants face a number of challenges from their surrounding environment. They encounter different stressful conditions, which may arise from living organisms (biotic stress), nonliving things (abiotic stress), or metals (heavy metal stress). To cope with the changing environment and optimize growth, plants have developed various strategies. The use of plant growth regulators as defense molecules has been the first choice for plants.

The major signaling pathways of plant growth regulators, namely jasmonates, salicylates, ethylene, strigolactones, and brassinosteroids, are known to play a part in various defense responses. Plant signal transduction shows extensive cross-talk among different signaling pathways, unlike the linear chain of events that was previously thought.

Jasmonates are members of the oxylipins family and are synthesized through the octadecanoid pathway. Jasmonates have been found to play a very important role in regulating a wide range of physiological and developmental processes, such as seed germination, promotion of root growth, stimulation of gravitropism, formation of trichomes, embryo development, seedling development, tuber formation, fruit ripening, leaf senescence, and alleviating the abiotic and biotic stress faced by plants. MeJA is a well-known organic compound that has been shown to be successful in preventing post-harvest fungal infections and extending the shelf life of fruits. The role of jasmonates in defense responses and growth protection promises to pave the way for the protection of crop plants and increase crop productivity, which ultimately leads to food security.

The present book provides a wealth of information about jasmonate structure, biosynthesis, signaling, practical applications of jasmonates in horticulture, their role in plant growth and development, cross-talk of jasmonates with secondary metabolites, antioxidants, and secondary messengers, their role in mediating stress responses, cross-talk of jasmonates with other phytohormones during abiotic stress tolerance, and the protective role of jasmonates during heavy metal stress.

CHAPTER 1

Practical Applications of Jasmonates in Agriculture

A. PRAVIN KUMAR, IQRA FAROOQ, and SHAHID RASOOL

CSIR–Indian Institute of Integrative Medicine, Field Station, Pulwama, Jammu and Kashmir, India

ABSTRACT

Endogenous plant growth characteristics are influenced by biotic and abiotic factors and events that play a role in its regulation and development. A specific category of growth regulators known as "jasmonates," that cause various reactions in plants and define the response of plants to the stressful events. Jasmonates diversely influence plant growth, including the development of storage organs, activation of plant defense mechanisms against biotic and abiotic stresses and growth inhibition in tissues like roots and young shoots. Plants under stress conditions produce secondary metabolites that could be used in the development of functional foods, nutraceuticals, therapeutics, cosmetics, and agrochemicals. Jasmonates have an ability to interact with other hormonal pathways, especially ethylene to have a significant impact in plant growth and development and as such can be optimally utilized for quality yields in field and horticultural crops. Jasmonates enhances the defense capability of the plant against biotic stresses like pathogen, viral, and insect pest infestation; and abiotic stresses like water stress, heat, and cold stress and salinity stress. It also improves the quality and nutritional longevity of the crop produce through enhanced shelf life. This chapter discusses the influence of Jasmonates on growth and development, physiological processes, defense

Biotechnological Applications of Jasmonates: A Group of Less Well-Known Plant Growth Regulators. Tariq Ahmad Dar, Bilal Ahmad, and Khalid Rehman Hakeem (Eds.)

mechanisms of plants and increased and qualitatively better yields in field and horticultural crops.

1.1 INTRODUCTION

Due to its role in growth and development as well as the defense mechanisms against various biotic and abiotic stresses, jasmonates, a category of phytohormones formed through the breakdown of membrane fatty acids, have attracted a lot of interest for its application in crop production. Jasmonates are class of crucial signaling chemicals that includes Jasmonic acid, Methyl jasmonates, and the intermediate hormones that occur in the octadecanoid pathway. The series of metabolic events that lead to the synthesis of jasmonates when linolenic acid is oxidized is known as the octadecanoic pathway (Ketabchi and Shahrtash, 2011). The chemicals or signaling reactions that allow plants to respond to stresses are regulated by these metabolites. In a prominent position, phyto-hormones serve significant regulatory roles in plant physiology (Chen et al., 2006; Browse, 2009). Jasmonic acid and methyl jasmonates (MeJA), which are naturally occurring plant growth regulators, are found throughout the plant kingdom and regulate a number of elements of plant development and just how plants respond to environmental stresses (Meyer et al., 2003). MeJA has the ability to influence a variety of morphological, physiological, and biochemical processes in plants (Norastehnia et al., 2007; Saniewski et al., 2013). It is typically thought to regulate the opening and closing of stomata, cell division, plant growth, and photosynthetic processes (Anjum et al., 2011). Chemical elicitors are substances that can raise the phenolic content of fruit (Ruiz-García and Gómez-Plaza, 2013). Elicitors are chemicals that can activate a plant's defense mechanisms, so enhancing the plant's resilience to pathogen attacks (Delaunois et al., 2014). One of the indubitable defensive reactions is the activation of secondary metabolic pathways. An effective antioxidative system is frequently associated with the reduction of oxidative damage and an increase in resilience to environmental challenges throughout critical growth stages of plants. Chemical elicitors such as jasmonic acid (JA) and in particular, its methyl ester, methyl jasmonate (MeJA), are regarded to be molecules of interest for triggering plant defenses and so improving quality of crop produce (Rohwer and Erwin, 2008). MeJA applied via the foliar route alters a number of physiological responses, increasing resistance to abiotic stressors (Walia et al., 2007). MeJA's function in defending plants from diverse abiotic stressors has been debatable, though. For instance, it has been claimed that MeJA increases drought tolerance in rice (Ho et al., 2015). Jasmonates, when administered exogenously, cause a variety of physiological reactions to stress, increasing plant resilience stressors (Walia et al., 2007; Ding et al., 2002). MeJA was previously categorized by the U.S. Food and Drug Administration (FDA) as a chemical that is generally recognized as safe (GRAS), and has scope for commercial applications in postharvest treatments to maintain the quality of different horticultural crops (Wang et al., 2010). Utilizing MeJA as a fumigant after harvest in various agricultural crops can activate defenses and prevent postharvest infections.

1.2 EFFECT OF JASMONATES IN ABIOTIC STRESS

Growing evidence suggests that numerous plant species undergoing salt stress respond favorably to jasmonates especially methyl jasmonate (MeJA) and jasmonic acid (JA) (Manan et al., 2016; Ahmadi et al., 2018). Most crops cannot withstand high salt concentrations, which significantly restricts the use of salt-affected soils for crop cultivation. Different salt-degraded soils make up about 6% of the world's land area and 23% of its cultivated fields. (FAO, 2005). Most non-halophytic plants are significantly constrained in their ability to grow by salt stress. Lipid derivatives called jasmonates have an important signaling role in many plant stress reactions and growth. Abiotic stressors such as UV, drought, heavy metal, ozone, cold, high temperatures, and salt stress are known to be lessened by these substances (Dar et al., 2015). In a study it was reported that environmental stressors like salt stress can severely restrict rapeseed yield. In fact, where the saline issue already exists, rapeseed productivity faces significant hurdles (Ashraf and McNeilly, 2004). Production of phytohormones like abscisic acid, salicylic acid, and jasmonates is one of the responsive plant responses to salt stress that may help to reduce salinity stress (Yoon et al., 2020). MeJA has discovered to increase anthocyanin concentration to further combat the oxidative stress brought on by decreased photosynthetic activity and altered chlorophyll content in *A. thaliana*. and the activities of the antioxidant enzymes catalase (CAT), superoxide dismutase (SOD), peroxidase (Px) and glutathione reductase

(GR) (Maciejewska and Kopcewicz, 2002). MeJA, promotes the production of secondary metabolites like ganoderic acid and boosts the activation of several defense genes as well as host resistance genes (Chen et al., 2006; Ren et al., 2010) Numerous studies have clearly demonstrated the beneficial effects of jasmonate therapies on various aspects of plant growth and development under different stress (Dar et al., 2015).

1.3 APPLICATION OF MEJA IN AGRICULTURAL CROPS

The concentration of a number of amino acids in the Tempranillo grapes was raised by the foliar spray of methyl jasmonate. Since phenylalanine is a precursor to many phenolic and aromatic chemicals, it is important to note that the phenylalanine level was significantly increased. As this elicitor treatment improved the nitrogen content of the grapes, methyl jasmonate foliar application might be employed to raise the quality of the grapes and the wine (Garde-Cerdán et al., 2018). The MeJA foliar treatment to the grapevines had no effect on the remaining components or the total amount of terpenoids in the berries. In terms of phenolic-related measures, MeJA foliar treatment was superior to control in terms of colour intensity and total anthocyanin content. (Portu et al., 2015). The foliar application of MeJA enhanced the amount of benzyl alcohol and 2-phenylethanol in proportion to benzenoids, favorably altering the total benzenoids content. With MeJA application grapes could able to synthesis these compounds (Garde-Cerdán et al., 2018). Application of chemical fungicide reduced the concentration of number of amino acids content on other hand MeJA boosted the level of aromatic amino acids (Oliva et al., 2011; Garde-Cerdán et al., 2014).

1.3.1 PRE-HARVEST APPLICATION

Depending on the crop, dose, and phenological stage, MeJA treatment prior to harvest can have a variety of effects on plants. Pre-harvest treatments on Pharbitis nil (Japanese Morning Glory) had growth-suppressing effects similar to those of abscisic acid applications on the leaves, roots, buds, and shoots. However, administration of gibberellic acid largely corrected these effects (Reyes-Díaz et al., 2016). The concentration of significant health-promoting substances like ellagic acid, quercetin, and myricetin increased noticeably when MeJA was applied to raspberry plants prior to harvest (Li et al., 2010). After pre-harvest MeJA treatments, the antioxidant activity of *Myrica rubra* (Chinese bayberry) and *Lactuca sativa* (Lettuce) both elevated. As MeJA treatments resulted in increased phenylalanine ammonia-lyase (PAL) activity, the total phenolic content of both plant species elevated (Jung, 2004). Since rosmarinic acid and caffeic acid are two of this species' powerful antioxidant components, Kim et al. (2006) found that the total phenolic content of sweet basil increased significantly following the pre-harvest application of MeJA in comparison to the control. The researchers concluded that MeJA's stimulation effect on PAL enzyme activity was the reason for this increase (Li et al., 2010). Pre-harvest MeJA treatments may therefore be a useful strategy to improve fruit quality and lengthen shelf life, though the ideal hormone concentration will vary depending on the species and cultivar.

1.3.2 POST-HARVEST APPLICATION

The firmness, crispness, and juiciness of fruits' textures, as well as their hardness, color, and presence or absence of physiological illnesses or decay, as well as their phytosanitary status, have historically been important criteria in predicting how long they will keep after being harvested. Horticultural research has made advances in crop breeding, social customs, and post-harvest handling and storage technology as a result of post-harvest fruit traits. The main factors that could lower fruit quality at this time have been fruit shelf life and fungal infection (Shafiq et al., 2012). Although the use of synthetic chemical fungicides is the most effective strategy for reducing the frequency of diseases in fruits, effective and non-toxic alternatives must be developed to address this issue during harvest. Therefore, using natural hormones like MeJA is a preferred alternative to increase the prevalence of diseases in fruits. Exogenous MeJA treatments have been demonstrated to enhance fruit's postharvest disease resistance and decrease fungal attack, allowing for a longer and better postharvest life (Flores and Ruiz del Castillo, 2014). The effect of post-harvest application of MeJA for inhibition of fungal disease in some horticultural crops has been presented in Table 1.1. Residual soluble protein levels were 36.3% and 50.1% of initial levels in MeJA-treated mushroom fruit bodies (100 M and 10 M) after 7 days of storage, compared to 24.9% in untreated controls, indicating that the right concentration of MeJA may help maintain mushroom quality during postharvest storage. In plants, flavonoids function as light filters, photoreceptors, visual attractants, antimicrobials, antioxidants, and antimicrobials. A treatment with 100 μM MeJA significantly enhanced the amount of total flavonoids and kept that level higher than untreated controls for the first five days of storage. To maintain the total flavonoids content over this storage time, 10 μM MeJA treatment was less effective than 100 μM MeJA administration. However, on day 5, the total flavonoid content in the 10 μM MeJA-treated samples began to increase and by the end of storage, it was 1.17 times higher than that of the control. When compared to the control, *A. bisporus* fruit bodies treated with MeJA had significantly higher levels of total phenolics and flavonoids, which suggests that MeJA may improve the mushroom's antioxidant status by promoting phenolic metabolism (Meng et al., 2012).

TABLE 1.1 MeJA Effect on Post-Harvest and Storage Fungal Diseases

Crop	Application Method	Fungal Species	References
Grapevine	Vapor spray	*Erysiphe necator* *Botrytis cinerea*	Belhadj et al. (2006)
Papaya	Vapor	*Colletotrichum gloeosporioides*	Wang et al. (2003)
Tomato	Dipping	*Botrytis cinerea*	Zhu and Tian (2012)
Mandarins	Vapor	*Penicillium digitatum*	Guo et al. (2014)
Peach	Vapor	*Monilinia fructicola* *Penicillium expansum*	Yao and Tian (2009)
Strawberry	Vapor	*Epiphyas postvittana*	Ayala-Zavala et al. (2005)
Pear	Vapor	*Penicillium expansum*	Zhang et al. (2009)
Sweet cherry	Spray	*Monilinia fructicola*	Yao and Tian (2005)

1.4 HEALTHCARE COMPONENTS

Jasmonates and its derivatives may have both direct and indirect effects on human health. MeJA treatments given before and after harvest have been proven to promote the production of naturally occurring chemicals with advantageous properties in some plant species, boosting those plants' favorable health effects on humans (Flores and Ruiz del Castillo, 2014). Sayyari et al. (2011) demonstrated that MeJA treatments increase the bioactive components in pomegranates and blackberries (Wang et al., 2007) encouraging their qualities that are good for human health. Application of MeJA increases the antioxidant activity and other anti-oxidant enzymes like Ascorbate peroxidase, catalase, ascorbate peroxidase, polyphenol oxidase, PAL, flavanone 3-hydroxylase, and 1-aminocyclo-propane in various fruit crops when it applied in vapor form. Table 1.2 details about the crops showing higher antioxidant content following the application of MeJA. Mangoes, Kiwis, Radishes, and freshly cut fruits, vegetables treated with MeJA postharvest kept higher levels of sugars and organic acids, while blackberries, raspberries, and strawberries sustained greater amounts of bioactive components and demonstrated increased antioxidant potential (González-Aguilar et al., 2000; Wang & Buta, 2003; Wang, 2005; Chanjirakul et al., 2006; Chanjirakul et al., 2007; Palonen et al., 2021).

TABLE 1.2 Increase in Antioxidant Content in Fruits After Treatment with MeJA

Crop	Application Method	Enzymatic and Non-Enzymatic	References
Pomegranates	Vapor	Total Phenolis, anthocyanin	Sayyari et al. (2011)
Grapes	Vapor	Total Phenolis, anthocyanin	Flores et al. (2015)
Apple	Dipping	Anthocyanin	Rudell et al. (2005)
Plum	Vapor	ACS and ACO	Khana and Singha (2007)
Peach	Vapor	Sucrose phosphate synthase	Yu et al. (2016)
Strawberry	Vapor	Anthocyanin, phenolic acid, CAT, POD, PPO	Ayala-Zavala et al. (2005)
Raspberry	Vapor	PAL, anthocyanin, flavanone	Flores et al. (2014)
Blueberry	Spray	Anthocyanin and phenolic acid	Huang et al. (2015)

1.5 CONCLUSION

Jasmonates The formation of secondary metabolites in plants, such as anthocyanins, flavonoids, phenolic acids, and other antioxidant molecules, can be increased by jasmonates and their derivatives, which improves fruit quality, post-harvest life, and health benefits for people. MeJA is a well-known organic compound that has been shown to be successful in avoiding post-harvest fungus infections and extending the shelf life of fruits. Future studies on the usage of Jasmonates in combination with other beneficial components should be conducted in order for their synergistic impact to generate more healthy fruits. Identification of the genes that are activated and silenced in response to treatments in plant systems and it is also essential for understanding MeJA's action mechanism need to be studied and explored to a greater level.

KEYWORDS

- **abiotic stress**
- **antioxidant**
- **biotic stress**
- **defense mechanisms**
- **jasmonates**
- **methyl jasmonate**
- **phytochemicals**
- **secondary metabolites**

REFERENCES

Ahmadi, F. I., Karimi, K., & Struik, P. C. (2018). Effect of exogenous application of methyl jasmonate on physiological and biochemical characteristics of *Brassica napus* L. cv. Talaye under salinity stress. *South African Journal of Botany, 115*, 5–11.

Anjum, S. A., Wang, L. C., Farooq, M., Hussain, M., Xue, L. L., & Zou, C. M. (2011). Brassinolide application improves the drought tolerance in maize through modulation of enzymatic antioxidants and leaf gas exchange. *Journal of Agronomy and Crop Science, 197*(3), 177–185.

Ashraf, M., & McNeilly, T. (2004). Salinity tolerance in *Brassica* oilseeds. *Critical Reviews in Plant Sciences, 23*(2), 157–174.

Ayala-Zavala, J. F., Wang, S. Y., Wang, C. Y., & González-Aguilar, G. A. (2005). Methyl jasmonate in conjunction with ethanol treatment increases antioxidant capacity, volatile compounds, and postharvest life of strawberry fruit. *European Food Research and Technology, 221*, 731–738.

Belhadj, A., Saigne, C., Telef, N., Cluzet, S., Bouscaut, J., Corio-Costet, M. F., & Mérillon, J. M. (2006). Methyl jasmonate induces defense responses in grapevine and triggers protection against *Erysiphe necator*. *Journal of Agricultural and Food Chemistry, 54*(24), 9119–9125.

Browse, J. (2009). Jasmonate passes muster: A receptor and targets for the defense hormone. *Annual Review of Plant Biology, 60*, 183–205.

Chanjirakul, K., Wang, S. Y., Wang, C. Y., & Siriphanich, J. (2006). Effect of natural volatile compounds on antioxidant capacity and antioxidant enzymes in raspberries. *Postharvest Biology and Technology, 40*(2), 106–115.
Chanjirakul, K., Wang, S. Y., Wang, C. Y., & Siriphanich, J. (2007). Natural volatile treatments increase free-radical scavenging capacity of strawberries and blackberries. *Journal of the Science of Food and Agriculture, 87*(8), 1463–1472.
Chen, H., Jones, A. D., & Howe, G. A. (2006). Constitutive activation of the jasmonate signaling pathway enhances the production of secondary metabolites in tomato. *FEBS Letters, 580*(11), 2540–2546.
Dar, T. A., Uddin, M., Khan, M. M. A., Hakeem, K. R., & Jaleel, H. (2015). Jasmonates counter plant stress: A review. *Environmental and Experimental Botany, 115*, 49–57.
Delaunois, B., Farace, G., Jeandet, P., Clément, C., Baillieul, F., Dorey, S., & Cordelier, S. (2014). Elicitors as an alternative strategy to pesticides in grapevine? Current knowledge on their mode of action from controlled conditions to vineyard. *Environmental Science and Pollution Research, 21*, 4837–4846.
FAO, A. (2005). Global network on integrated soil management for sustainable use of salt-affected soils. *FAO Land and Plant Nutrition Management Service Rome.*
Flores, G., & del Castillo, M. L. R. (2014). Influence of preharvest and postharvest methyl jasmonate treatments on flavonoid content and metabolomic enzymes in red raspberry. *Postharvest Biology and Technology, 97*, 77–82.
Flores, G., Blanch, G. P., & del Castillo, M. L. R. (2015). Postharvest treatment with (−) and (+)-methyl jasmonate stimulates anthocyanin accumulation in grapes. *LWT–Food Science and Technology, 62*(1), 807–812.
Garde-Cerdán, T., López, R., Portu, J., González-Arenzana, L., López-Alfaro, I., & Santamaría, P. (2014). Study of the effects of proline, phenylalanine, and urea foliar application to Tempranillo vineyards on grape amino acid content. Comparison with commercial nitrogen fertilizers. *Food Chemistry, 163*, 136–141.
Gonzalez-Aguilar, G. A., Buta, J. G., & Wang, C. Y. (2003). Methyl jasmonate and modified atmosphere packaging (MAP) reduce decay and maintain postharvest quality of papaya 'Sunrise.' *Postharvest Biology and Technology, 28*(3), 361–370.
Guo, J., Fang, W., Lu, H., Zhu, R., Lu, L., Zheng, X., & Yu, T. (2014). Inhibition of green mold disease in mandarins by preventive applications of methyl jasmonate and antagonistic yeast *Cryptococcus laurentii. Postharvest Biology and Technology, 88*, 72–78.
Ho, J., Maeng, S., & Park, W. J. (2015). Effects of ethylene precursor, auxin, and methyl jasmonate on the aerenchyma formation in the primary root of maize (*Zea mays*). *Journal of Life Science, 25*(1), 37–43.
Huang, X., Li, J., Shang, H., & Meng, X. (2015). Effect of methyl jasmonate on the anthocyanin content and antioxidant activity of blueberries during cold storage. *Journal of the Science of Food and Agriculture, 95*(2), 337–343.
Jung, S. (2004). Effect of chlorophyll reduction in *Arabidopsis thaliana* by methyl jasmonate or norflurazon on antioxidant systems. *Plant Physiology and Biochemistry, 42*(3), 225–231.
Ketabchi, S., & Shahrtash, M. (2011). Effects of methyl jasmonate and cytokinin on biochemical responses of maize seedlings infected by *Fusarium moniliforme*. *Asian Journal of Experimental Biological Science, 2*(2), 299–305.

Kim, H. J., Chen, F., Wang, X., & Rajapakse, N. C. (2006). Effect of methyl jasmonate on secondary metabolites of sweet basil (*Ocimum basilicum* L.). *Journal of Agricultural and Food Chemistry, 54*(6), 2327–2332.

Li, Z., Hao, Y., Yang, Y., & Deng, W. (2010). Molecular cloning and expression analysis of a cytochrome P450 gene in tomato. *Plant Growth Regulation, 61*, 297–304.

Manan, A., Ayyub, C. M., Pervez, M. A., & Ahmad, R. (2016). Methyl jasmonate brings about resistance against salinity stressed tomato plants by altering biochemical and physiological processes. *Pakistan Journal of Agricultural Sciences, 53*(1).

Meng, D., Song, T., Shen, L., Zhang, X., & Sheng, J. (2012). Postharvest application of methyl jasmonate for improving quality retention of *Agaricus bisporus* fruit bodies. *Journal of Agricultural and Food Chemistry, 60*(23), 6056–6062.

Meyer, R., Rautenbach, G. F., & Dubery, I. A. (2003). Identification and quantification of methyl jasmonate in leaf volatiles of *Arabidopsis thaliana* using solid-phase microextraction in combination with gas chromatography and mass spectrometry. *Phytochemical Analysis: An International Journal of Plant Chemical and Biochemical Techniques, 14*(3), 155–159.

Norastehnia, A., Sajedi, R. H., & Nojavan-Asghari, M. (2007). Inhibitory effects of methyl jasmonate on seed germination in maize (*Zea mays*): Effect on α-amylase activity and ethylene production. *General and Applied Plant Physiology, 33*(1, 2), 13–23.

Oliva, J., Garde-Cerdán, T., Martínez-Gil, A. M., Salinas, M. R., & Barba, A. (2011). Fungicide effects on ammonium and amino acids of Monastrell grapes. *Food Chemistry, 129*(4), 1676–1680.

Palonen, P., Laine, T., & Mouhu, K. (2021). Floricane yield and berry quality of seven primocane red raspberry (*Rubus idaeus* L.) cultivars. *Scientia Horticulturae, 285*, 110201.

Portu, J., Santamaría, P., López-Alfaro, I., López, R., & Garde-Cerdan, T. (2015). Methyl jasmonate foliar application to Tempranillo vineyard improved grape and wine phenolic content. *Journal of Agricultural and Food Chemistry, 63*(8), 2328–2337.

Ren, A., Qin, L., Shi, L., Dong, X., Li, Y. X., & Zhao, M. W. (2010). Methyl jasmonate induces ganoderic acid biosynthesis in the basidiomycetous fungus *Ganoderma lucidum*. *Bioresource Technology, 101*(17), 6785–6790.

Reyes-Díaz, M., Lobos, T., Cardemil, L., Nunes-Nesi, A., Retamales, J., Jaakola, L., & Ribera-Fonseca, A. (2016). Methyl jasmonate: An alternative for improving the quality and health properties of fresh fruits. *Molecules, 21*(6), 567.

Rohwer, C. L., & Erwin, J. E. (2008). Horticultural applications of jasmonates. *The Journal of Horticultural Science and Biotechnology, 83*(3), 283–304.

Rudell, D. R., Fellman, J. K., & Mattheis, J. P. (2005). Preharvest application of methyl jasmonate to 'Fuji' apples enhances red coloration and affects fruit size, splitting, and bitter pit incidence. *HortScience, 40*(6), 1760–1762.

Ruiz-García, Y., & Gómez-Plaza, E. (2013). Elicitors: A tool for improving fruit phenolic content. *Agriculture, 3*(1), 33–52.

Saniewski, M., Czapski, J., & Horbowicz, M. (1994). Fatty acid and sterol contents during tulip leaf senescence induced by methyl jasmonate. *Acta Agrobotanica, 47*(2), 89–95.

Sayyari, M., Babalar, M., Kalantari, S., Martínez-Romero, D., Guillén, F., Serrano, M., & Valero, D. (2011). Vapor treatments with methyl salicylate or methyl jasmonate alleviated chilling injury and enhanced antioxidant potential during postharvest storage of pomegranates. *Food Chemistry, 124*(3), 964–970.

Shafiq, M., Singh, Z., & Khan, A. S. (2013). Time of methyl jasmonate application influences the development of 'Cripps Pink' apple fruit color. *Journal of the Science of Food and Agriculture, 93*(3), 611–618.

Walia, H., Wilson, C., Condamine, P., Liu, X., Ismail, A. M., & Close, T. J. (2007). Large-scale expression profiling and physiological characterization of jasmonic acid-mediated adaptation of barley to salinity stress. *Plant, Cell & Environment, 30*(4), 410–421.

Wang, C. Y., & Buta, J. G. (2003). Maintaining quality of fresh-cut kiwifruit with volatile compounds. *Postharvest Biology and Technology, 28*(1), 181–186.

Wang, K., Jin, P., Shang, H., & Zheng, Y. (2010). Effect of methyl jasmonate in combination with ethanol treatment on postharvest decay and antioxidant capacity in Chinese bayberries. *Journal of Agricultural and Food Chemistry, 58*(17), 9597–9604.

Wang, S. Y., & Zheng, W. (2005). Preharvest application of methyl jasmonate increases fruit quality and antioxidant capacity in raspberries. *International Journal of Food Science & Technology, 40*(2), 187–195.

Yao, H. J., & Tian, S. P. (2005). Effects of a biocontrol agent and methyl jasmonate on postharvest diseases of peach fruit and the possible mechanisms involved. *Journal of Applied Microbiology, 98*(4), 941–950.

Yoon, Y., Seo, D. H., Shin, H., Kim, H. J., Kim, C. M., & Jang, G. (2020). The role of stress-responsive transcription factors in modulating abiotic stress tolerance in plants. *Agronomy, 10*(6), 788.

Yu, L., Liu, H., Shao, X., Yu, F., Wei, Y., Ni, Z., & Wang, H. (2016). Effects of hot air and methyl jasmonate treatment on the metabolism of soluble sugars in peach fruit during cold storage. *Postharvest Biology and Technology, 113*, 8–16.

Zhang, H., Ma, L., Turner, M., Xu, H., Dong, Y., & Jiang, S. (2009). Methyl jasmonate enhances biocontrol efficacy of *Rhodotorula glutinis* to postharvest blue mold decay of pears. *Food Chemistry, 117*(4), 621–626.

Zhu, Z., & Tian, S. (2012). Resistant responses of tomato fruit treated with exogenous methyl jasmonate to *Botrytis cinerea* infection. *Scientia Horticulturae, 142*, 38–43.

CHAPTER 2

Fragrant Signals (Jasmonates) in Growth and Development

MD. ADIL,[1] UMAR MAJEED WANI,[1] ASIF AMIN,[1] FAIZAH FAROOQ,[1] VASEEM RAJA,[2] and RAIES A. QADRI[1]

[1]*Department of Biotechnology, University of Kashmir, Srinagar, Jammu and Kashmir, India*

[2]*University Center for Research and Development, Chandigarh University, Mohali, Punjab, India*

ABSTRACT

Jasmonates are a group of plant hormones that play important roles in regulating growth, development, and stress responses. Jasmonates belong to a family of phytohormones that is involved in different aspects of plants growth and development. Derived from linolenic acid in the chloroplast, in the initial steps of its biosynthesis linolenic acid gets converted into (OPDA) 12-oxophytodienoic acid (OPDA). OPDA is converted to jasmonic acid and its derivatives such as methyl Jasmonate (MeJA), Jasmonoyl-isoleucine (JA-Ile) and cis-Jasmone. They are derived from the fatty acid linolenic acid and are found in all higher plants. Jasmonates are involved in many processes such as root growth, flower development, and fruit ripening. They also play a role in plant defense against herbivores and pathogens, by triggering the production of defensive compounds such as terpenoids and phenolics. Jasmonates are involved in regulating the development of pollen and ovules, as well as in the synthesis of floral scents that attract pollinators. Additionally, jasmonates help plants adapt

Biotechnological Applications of Jasmonates: A Group of Less Well-Known Plant Growth Regulators. Tariq Ahmad Dar, Bilal Ahmad, and Khalid Rehman Hakeem (Eds.)

to environmental stresses by regulating the expression of stress-responsive genes and by modulating the synthesis of protective compounds. The signaling pathways of jasmonates are complex and involve many different components, making them a rich area of study for plant biologists. Understanding the role of jasmonates in growth and development can have important implications for improving crop yield and enhancing plant defense against stressors.

2.1 INTRODUCTION

Jasmonates are oxylipins (oxygenated lipids) that form a class of growth regulators in plants they are classified as type of sesquiterpenes, they include jasmonic acid and its derivatives. Originally discovered from the essential oils of jasmine in the 1960s. Jasmonates biosynthesis begins from polyunsaturated fatty acids obtained from the membranes of chloroplast through the octadecanoid pathway. They are majorly associated with plant defense mechanisms, they were initially considered to be associated suppression of growth and senescence (Murofushi et al. 1999). The inhibitory effect on growth by JA is crucial for the survival of the plants but over the years their role in different growth and developmental processes has been well documented.

2.2 FLORAL DEVELOPMENT AND JASMONATES

The floral induction or initiation in underpinned to the expression of certain genes called the floral integrators, such as FLOWERING LOCUS T (FT), SUPPRESSOR OF CONSTANS1 (SOC1) and LEAFY (LFY), etc. All the endogenous and exogenous cues (hormonal, photoperiod, temperature, aging, and vernalization) converge to the expression of these genes which determine the shoot apical meristem to floral identity (Kinoshita & Richter, 2020). The development of flower into different whorls is specified by activities of three class of genes called the A, B, and C genes. these homeotic genes that encode MADS-domain transcription factors like APETELLA1, APETELLA2 (class A genes), APETELLA3, PISTILLATA (class B), and AGAMOUS (AG), and so forth. Mutation in these genes leads to defective phenotypes in floral development (Irish, 2017).

The role of Jasmonate signaling is crucial in floral initiation and development. JAs regulate flowering via cross-talk with other signaling pathways. Defects in JA biosynthesis, perception, and signaling render the plants male sterile (Acosta & Przybyl, 2019; Browse & Wallis, 2019). *fad3 fad7 fad8, lox3 lox4, aos/dde2, opr3/dde1, acx1 acx5*, and *jar1-11 gh3.10-2* mutants defective in JA biosynthesis are male sterile, and show improper stamen development delayed anther dehiscence and unviable pollen grains in Arabidopsis (Caldelari et al., 2011). However, application of JA exogenously rescues stamen development and fertility in *opr3* mutants. Mutants defective in jar1 (enzyme found in JA-Ile synthesis) and jat1 (JA transporter) exhibit defective and late stamen development (Li et al., 2017). Similar defects are seen in JA receptor (COI1) mutants such as *coi1-1 (W467*), coi1-2 (L245F),* and *coi1-4 (G369E*) (Feys et al., 1994; Huang et al., 2014; Xie et al., 1998). over expression of JAZ proteins (JAZ1Δ3A or JAZ10.4) with deletion at of C-terminal JAS domain leads to abnormal development of stamens (Chung & Howe, 2009; Thines, et al., 2007). However, higher order mutation of strong JAZ variants hampers fertility, normal development, indicating that optimum levels of JAZ are required for appropriate stamen growth and development (Guo, et al., 2018). There are 12 JAZ proteins known in Arabidopsis till date. They interact with Jasmonate responsive transcription factors (MYC) and their complexes with other transcription factors to suppress their activity. MYB24 form a complex with bHLH transcription factor MYC2/3/4/5 (MYB-bHLH complexes) which cooperatively regulate stamen development and seed production (Qi, et al., 2015). MYB21/24 induce MYB108 expression in flowers and in mutants with defective MYB108 exhibit reproductive developmental abnormalities (Mandaokar & Browse, 2009). *myb21* and *myb24* mutants show similar disabilities in fertility and stamen development (Huang et al., 2017; Song et al., 2011). Quadruple mutants with *myc2 myc3 myc4 myc5* show lower expression of MYB21, MYB24, MYB57 and MYB108 levels in the early stages of floral development indicating at their upregulation occurs as a result of Jasmonate signaling. These mutants showed similar defects like delayed stamen development and were partially infertile (Qi, et al., 2015). JA signaling influences petal size in Arabidopsis, in comparison to the wild type, mutants defective in JA synthesis/perception show smaller petal size at the early stages of anthesis, that are corelated to reduced levels of MYB21 in petals (Reeves et al., 2012). JAs up regulate the expression of transcription factor BIGPETALp at during the floral development to restrict petal expansion (Brioudes, et al., 2009).

In tomato silencing of JA synthesis genes such as *SPR2*, *SLAOC* renders the plants female infertile (Goetz, et al., 2012; Li, et al., 2003). The ortholog of Arabidopsis COI1 in *Solanum lycospermum* jasmonic acid insesitive1-1 (jai-1) show reduction in pollen viability and female infertility and premature stamen dehiscence (Dobritzsch et al., 2015; Li et al., 2004), with abnormal pattern of JA biosynthesis and lowered SLMYB21 levels in flower buds few days before opening along with arrested flower development before opening (Dobritzsch et al., 2015; Niwa et al., 2018). Artificial repressor like AtMYB24-SRDX intended to inhibit SLMYB21 resulted in flower defects as in *jai1-1* mutants, unviable pollen grains, pistils incapable of hosting pollen tube elongation (Niwa et al., 2018).

Similarly, role of Jasmonates in flowering is conserved in other plants. Silencing of cotton acyl-COA-N-acyltransferase (GHACNAT) through virus induced gene silencing produced plants with indehiscent anthers and defective stamens and sterility (Fu et al., 2015) with reduced levels of JA (Fu et al., 2015). Anther dehiscence was rescued by exogenous application of methyl Jasmonate (Fu et al., 2015). CsGL2-like which encodes a HD-ZIP IV TF plays important role in development of male flowers in cucumber, silencing of this gene delayed male flowering and lowered pollen vigor (Cai et al., 2020). It interacts with JAZ (CsJAZ1) to form a complex (CsJAZ complex) (Cai et al., 2020). *Artesimia annua* COI1 overexpression improved fertility of Arabidopsis coi1-16 loss of function mutants (Cai et al., 2020). R2R3-MYB108 silenced capsicum plants showed increased capsaicinoids production and delayed anther dehiscence and decreased viability in pollen grains (Sun et al., 2019).

2.3 JASMONATES IN FINE TUNING FLORAL TIMING

The inhibitive role of JA in flowering timing is widely accepted, JA signaling delays floral transition in Arabidopsis via MYC2, MYC3, MYC4 transcription factors. JAZ repressors being a repressor of JA signaling promote flowering, JAZ proteins attenuate the JA signaling in absence of bioactive Jasmonates, they prevent the activity of Jasmonate responsive transcription factors by physically interacting with them. In absence of bioactive JA, JAZ proteins activate expression of various flowering associated genes like FT (FLOWERING LOCUS T) by interacting with numerous transcription factors and TOE1 and TOE2 (TARGET OF EAT). Over expression of JAZ proteins in tomato resulted in early flowering. Transgenic lines carrying these double or triple MYC mutants show early flowering, which was associated with increased levels of *FT* and *TSF* genes. MYC3 also binds to the promoter of FT to supress its transcription and delay flowering. bHLH type IIId transcription factors bHLH3 13, 14 and 17 are known targets of JAZ proteins. These TFs repress the JA responses by acting as Transcriptional repressors. bHLH3 IIId 13,14,17 quadruple mutants show late flowering where as their overexpression show results in early flowering, these factors antagonize function of other transcription factors in the JA signaling. Although, JAZ repressors in Arabidopsis and *Solanum lycospermum* promote flowering but it is not the case in other plants where their homologous genes show different functions. Likewise, *CmJAZ1*-like gene lacking Jas domain delays flowering when overexpressed in *Chrysanthemum morifolium* (Figure 2.1).

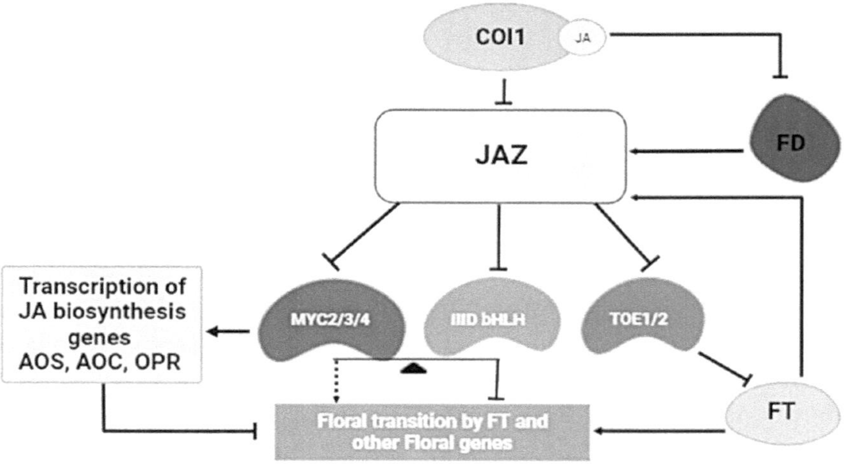

FIGURE 2.1 It represents the regulation of FLOWERING LOCUS T and other flowering genes by Jasmonate signaling. the arrows represent activation and the bars represent inhibition.

Abreviations: AOS: allene oxide synthase; AOC: allene oxide cyclase and FD a basic leucine zipper transcription factor that binds with FT in the florigen associated complex.

MYC and COI1 can interact with MED25 a component of plant Mediator complex (a multi-subunit transcriptional coactivator complex that binds to specific transcription factors, and RNA polymerase II to regulate various aspects of transcription) MED25 interaction with COI1 is more prominent but when JA in present COI1 forms COI1-JAZ complex to mediate JAZ degradation and MED25 becomes free to interact with MYC TFs. MED25- MYC2 this complex enhances the activity of MYC2 and the expression of target genes. This MED25 mediated enhancement brings early flowering by increasing LFY, FT, FUL. Despite enough evidence to support that Jasmonates delay floral induction, there are exceptions where the JA were reported to promote flowering. In oilseed rape, the exogenous treatment of MeJA leads to earlier flowering. In another case, JA treatment was able to enhance flowering, but only under low concentrations. This indicates that the role of JA in regulation of flowering is different in different plants (Zhao et al., 2022).

2.4 CROSS-TALK WITH GIBBERELLIN SIGNALING IN REGULATION OF FLOWERING

JA and GA show antagonistic relationship in regulating flowering. Where JA are known to delay flowering, GA promotes floral transition. JA levels is high in during the juvenile phase to favor vegetative growth over reproductive growth, simultaneously the levels of GA are low in the juvenile phase but begin to increase with aging, the high levels of GA are needed for the transition towards reproductive phase.

In cotton ectopic expression of Gh166 containing an alpha- amylase inhibitor (AAI) Induces early flowering, as well as reducing the transcription of JA signaling genes and upregulating the GA signaling genes (Figure 2.2). JA are considered to be active in Thigmomorphogenesis, during touch stimulus *GA2ox6* (a GA catabolic gene) is induced resulting in reduction of GA levels and delay in flowering. JA could be responsible for the decrease in GA levels to delay flowering in response to touch stimulus. Apart from this there is cross-talk between the components of JA and GA signaling. JAZ proteins interact with Della proteins and prevent inhibition of transcription factors like WRKY, CO, and PIFs. In presence of JA, JAZ proteins are degraded and this repression of DELLA proteins on GA signaling is reinforced. RGL3 a DELLA protein is targeted by MYC2 during and JA mediated responses such as delaying floral transition during touch response, to inhibit the GA signaling.

According to the "relief of repression" model proposed by Hou et al. (2010). DELLA proteins modulate JA signaling by binding to JAZ proteins in presence of JA, JAZ proteins are degraded and MYC2 is released to bring about change in gene expression. When GA is present the DELLA proteins are degraded, allowing JAZ proteins to be able to bind to MYC2 and other transcription factors to prevent their activity. Degradation of DELLA proteins allows for the production of FT and flowering. DELLA proteins also repress the transcription of *mi172* that regulates TOE1 and TOE2 which are activated by JA signaling and negatively regulate the expression of FT. The cross-talk between GA and JA signaling takes place primarily in the leaves, which is the site of FT formation and is important for the fine tuning of floral timing.

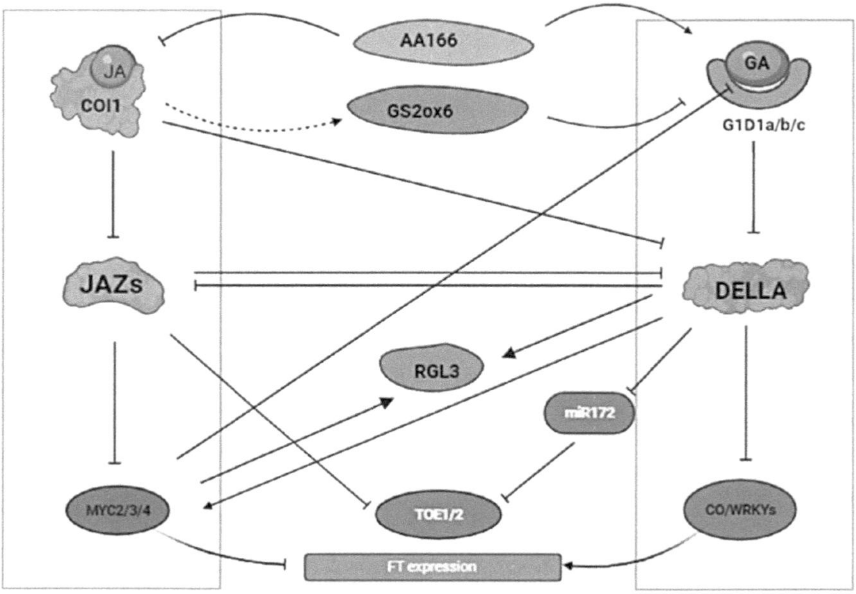

FIGURE 2.2 Representing cross-talk between JA and GA signaling in controlling flowering, JA and GA control flowering by antagonistically regulating Flowering locus T expression. broken lines indicate indirect regulation.

Source: Modified figure from Zhao et al., (2022).

The levels of JA and GA in different stages of plants are consistent with this model, in the juvenile phase of plants, the low levels of GA are high levels of DELLA proteins are present to prevent flowering. With time the levels of GA show a gradual increase, so do the JAZ proteins to prevent the inhibition of flowering. Indicating to an antagonistic relationship between JA and GA signaling regulation of flowering (Zhao et al., 2022).

2.5 JASMONATES IN PRODUCTION OF NECTAR

Nectar is a sugar-water solution that is produced by plants in their flowers, the purpose of producing nectar is to attract agents of cross pollination (insects and birds), as strategy to prioritize outcrossing which is essential introduce genetic diversity (Radhika et al., 2010). There are two types of nectars the floral nectar and the extra floral nectar, the latter is produced from glands (extra floral nectarines) but are not involved in pollination. They are a part of plants defense by indirectly by attracting plant friendly arthropods. The composition of nectar can vary in different plants, but carbohydrates (glucose and fructose), amino acids and water are main components of nectar and function to attract pollinators, other components such as Nectarins (nectar proteins with anti-microbial activity), various volatile compounds that impart odour to the nectars. Nectar can virtually be produced from any part of the plant except roots. They are produced from nectarines which can be diverse in terms of their position, mechanisms of secretion (Heil, 2011). Floral nectarines can be present in different parts of the flower, based on their position they have been divided into different types such as, gynopleural, stigmatic, staminal, hypanthial, perigonial, receptacular, petal, and sepal, and so forth (Nicolson, Nepi, & Pacini, 2007). They secrete nectar through epidermal tissue through trichomes or Stomata that remain always open (Heil, 2011). Extrafloral Nectarines are present outside flowers and are not involved in pollination. They are usually present on the leaf or petiole but have been found a range of different organs (Rudall, 2020).

2.6 REGULATION OF NECTAR BY JA

In *Brassica napus* during the course of floral development the there is a burst of endogenous JA level which is followed by the peak of nectar secretion indicating JA regulate the synthesis of nectar. Consistently the exogenous application of JA was able increase the nectar synthesis at the highest stage of nectar production, use of inhibitors like coronalon and phenidone at the early stages of flowering decreased the levels of nectar production and interfered with the development of flower (Radhika et al., 2010). In *Phaseolus lunatus* production of Extrafloral nectar was enhanced by light in JA- and JA-isoleucine treated plants application (Radhika et al., 2010b) (Figure 2.3).

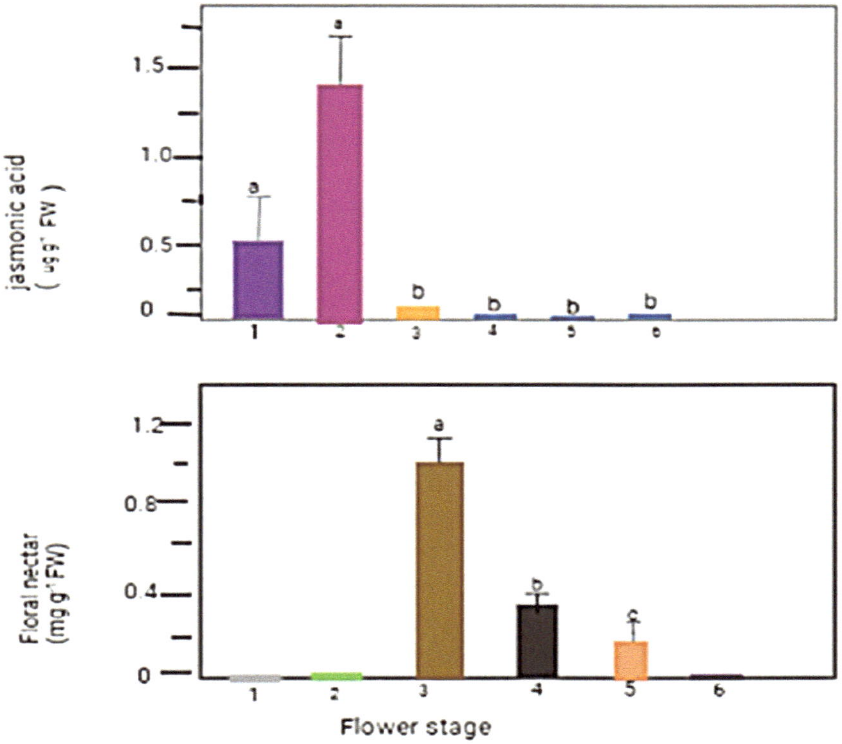

FIGURE 2.3 Represents the correlation between levelsl of Jasmonic acid and nectar secretion in brassica. As can be seen from the plots the initial burst of endogenous JA levels resulted in the secretion of nectar.

Source: Modified figure from Radhika et al., (2010a).

In *Nicotiana attenuata* silencing of JA biosynthetic gene AOC (ir-*aoc*), and COI1 (ir-*coi1*) led to a decrease of in the transcription of *SWEET9* (a sucrose transporter gene present in the nectary tissue) gene compared to controls, Benzyl acetone (pollinator attracting scent in *Nicotiana attenuata)* levels were extremely low in ir-*aoc*, due a decrease in the expression of *chalcone synthase 1. However*, the levels of BA in ir-*coi1* were similar to controls. In the flowers of both transgenic lines the levels of defensive secondary metabolites like nicotine, (E)-α-bergamotene, caffeoylputrescine, and so forth and defensive proteins like defensin, trypsin proteinase inhibitor were lower (Li et al., 2018). In *Nicotiana tabacum* RNA interference mediated silencing of NtCOI1 negatively affected the expression of *NtMYB305* which mediates starch metabolism in tobacco and led to loss of floral nectar (Wang et al., 2014). Arabidopsis flowers possess two types of Nectaries, median, and lateral nectaries. The median nectaries are present at the base of petals stamens, lateral nectaries are present at the base of short stamens. The JA biosynthesis mutants nectar production was not observed in young flowers, in aos-2 mutants there was no nectar produced at all. *AOS* (allene oxide synthase, a single copy gene) is indispensable for JA biosynthesis, in this case, it indicates the nectar production is dependent on octadecanoic pathway. But in case of *dad1-1* and *opr3-1* (neither of these two mutants were completely JA deficient) mutants the old flowers eventually develop nectar. Similarly, the young flowers of *coi1-1* mutants nectar developed only in old flowers and exogenous application of MeJA, and auxin treatment was able to rescue nectar production in young flowers of *coi1-1* mutants. The development of nectar in old flowers indicates the presence of an alternate COI1 independent route for nectar production. A Jasmonate responsive transcription factor MYB21 know to be involved in stamen development is regulates nectar production and by activating the expression of nectar specific genes like *AtSWEET9* (a sucrose transporter gene that transports sucrose from nectary parenchyma cells) and *AtCWINV4* (CWINV4 converts sucrose in hexose monomers). *myb21-24* mutants contained nectar less flowers with very low expression levels of *SWEET9*, *CWINV4*, and *PIN6*. *myb21-24* mutants were not rescued by exogenous MeJA treatment. Starch accumulation takes place in nectaries and this starch is converted into mono and disaccharides is needed for nectar secretion. In transgenic lines with nectar less flowers showed high starch staining patterns (Schmitt et al., 2018).

2.7 JASMONATES IN LEAF SENESCENCE

Leaf senescence is the last stage in the process of leaf development. In this stage age dependent disassembly and relocation of macromolecules and nutrients to other parts of plants takes place. Combined action of environmental and intrinsic signals results in induction of leaf senescence mediated by a regulated mechanism. There are two types of senescence that plants show, mitotic, and postmitotic senescence. Mitotic senescence refers to the arrest of replication in meristematic cells. Postmitotic senescence is the planned and controlled disassembly and degeneration of cellular content in plant organs that have differentiated and matured. The metabolic changes that occur inside the senescent organ is that the plant shifts from anabolism to catabolism. The first morphological feature of leaf senescence is yellowing due degradation of chlorophyll, that occurs via the PAO/phyllobilin pathway. The catabolism of chlorophyll is begins by the conversion of chlorophyll *b* to *a,* with help of chlorophyll b reductase encoded by NONYELLOW COLORING1 (NYC1) and NONCOLORING1-LIKE (NOL) and HCAR. And finally converted into primary fluorescent chlorophyll catabolite (*pFCC),* which subsequently gets turned into nonfluorescent products in the vacuole. Chloroplast degradation takes place with the hydrolysis of macromolecules like lipids with and proteins. Phospholipase D, lyticacylhydrolase, and lipoxygenase hydrolyse the chloroplast membrane lipids. Mitochondrial ROS act as a signal to nucleus to activate transcription factors like WRKY in leaf senescence. Leaf senescence is majorly controlled by two family of transcription factors WRKY and NAC. Morphological changes appear in the mitochondria, they become more round and enlarges during senescence [38, 39].

2.8 JASMONATES IN REGULATION OF LEAF SENESCENCE

Regulation of leaf senescence is a result of a combined effect of different intrinsic factors like hormones, age, nutrients, and extrinsic ques like temperature, light, and pathogens. JA are known to modulate senescence in plants. In senescent leaves of Arabidopsis, the levels of JA are 4 times higher than non-senescent leaves, also high levels of JA associated genes like *LOX 1,2,3* (lipoxygenase), *AOS* (allene oxide synthase), *AOC1* (allene oxide cyclase1) are present. Applying MeJA exogenously induces leaf senescence in detached and attached leaves. Incubation of expanding leaves with MeJA for longer duration induces senescence with morphological symptoms like browning. MeJA induces accumulation of SAG12, a marker for leaf senescence.

TEOSINTE BRANCHED/CYCLOIDEA/PCF (TCP) family of transcription factors controlled by *miRNA319*, activate *LOX2* during leaf senescence. *miRNA319* over expressing mutants show delayed senescence, but use of MeJA restores senescence. JA insensitive coi1 mutant plants do not display symptoms of leaf senescence in presence of JA. Rubisco activase (RCA) a JA-COI1 responsive proteins that is down regulated in JA signaling. Out of 12 JAZ proteins in Arabidopsis JAZ 8 and JAZ4 participate in leaf senescence, *jaz4* and *jaz8* mutants show senescence phenotypes but their overexpression results in delayed senescence under the treatment of MeJA. JAZ7 is required in delaying dark induced senescence. The JA signaling transcription factors MYC2, MYC3, and MYC4 bring transcriptional changes and activate key genes like Pheophorbide A Oxygenase (PAO) which is responsible for chlorophyll degradation. *bHLH3*, *bHLH13*, *bHLH14*, and *bHLH17* act antagonistic to MYC2, MYC3 MYC4 by attenuating the JA mediated leaf senescence [38–40]. In tomato over expression of Sl JAZ10 and SlJAZ11 showed negatively influenced dark induced leaf senescence, over expression of these proteins led to decreased levels of Age related genes and resistance to dark induced leaf senescence (Tang et al., 2022). In tomato, RNA1 silencing of SlMYC2 demonstrated that SlMYC2 activates expression of chlorophyll degrading enzyme S*l*PAO, and inhibited the expression of S*l*RCA and S*l*BPASE (genes involved in photosynthesis) (Ding et al., 2022) (Figure 2.4).

2.9 CROSS-TALK WITH OTHER HORMONE SIGNALING PATHWAYS

Salicylic acid, another plant hormone that contributes significantly to the regulation of senescence. The levels of SA increases with time to eventually induce the expression of SAGs and bring about senescence. JA and SA regulate leaf senescence antagonistically, by regulating the senescence specific transcription factor *WRKY53*. This TF is induced by SA and interacts with EPITHIOSPECIFYING SENESCENCE REGULATOR in the nucleus which itself is induced by JA-signalling. due to binding the activity of WRKY53 and impacts its ability to bind target regions in the genome.

EIN2 ETHYLENE INSENSITIVE 2 (*EIN2*) a component of Ethylene signaling is critical to leaf senescence. *EIN2-1* mutants application of

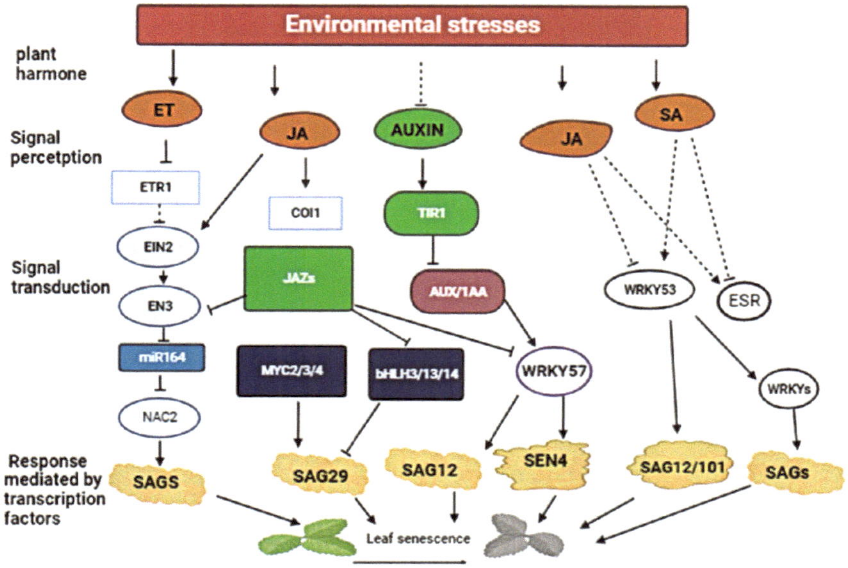

FIGURE 2.4 Regulation of leaf senescence in mediated by different hormone signaling and their cross-talk. dotted lines represent indirect regulation, arrows bars represent activation and inhibition. Abbreviations: TIR! Transport inhibitor resonse1, ETR1, ethylene response1, SA, salicylic acid, ESR, epithio specifying senescence regulator, SAG, senescence associated genes.

Source: Modified figure from Hu et al., (2017).

ethylene is unable to induce floral abscission. *EIN3* mutants display delayed leaf senescence, *EIN3* mutant show a delayed JA induced leaf senescence. *EIN3* represses miRNA164 and upregulates its target *NAC2* transcription factor to regulate leaf senescence.

Auxin is a suppressor of leaf senescence, with age the levels of auxin decreases in leaves, JA and AUX signaling cross-talk through *WRKY57* which is induced by AUX and inhibited by JA. *WRKY57* mutants displayed symptoms of leaf senescence, with high expression of SAGs. *WRKY57* represses the transcription of *SENESCENCE4*, *SENESCENCE-ASSOCIATED GENE12*. *WRKY57* Also interacts with repressors of JA and AUX signaling, JAZ4 and JAZ8 proteins and IAA29. WRKY57 could be a repressor of the JA mediated leaf senescence (Hu et al., 2017; Jiang et al., 2014; Woo et al., 2019).

2.10 ROOT GROWTH AND JASMONATES

The root tip consists of three regions, the meristematic, elongation, and differentiation zones. The meristematic region contains the stem cell niche. Stem cells are present in quiescent center, its maintenance is necessary for the growth and elongation of root. *Wox5* a transcription factor (WUSCHEL-LIKE HOMEOBOX5) is a root stem cell organizer that maintains the activity of quiescent center. It prevents the proliferation of quiescent center by suppressing CYCD3;3 expression. *WOX5* moves to adjacent cells where it suppresses their differentiation. In columella stem cells suppression of differentiation is mediated by TOPLESS-HISTONE DEACETYLASE 19. PLETHORA (PLT) transcription factors regulate root organogenesis by regulation of proliferation and suppress differentiation, mediate auxin effects. overexpression of PLT genes results in abnormal root development. Across the meristem it exists as a gradient with maximum in the stem cells. PLT suppress GROWTH REGULATING FACTOR (which inhibits the PLT itself) through miRNA396. GRAS transcription factor SCARECROW (SCR) regulates cortex-endodermis initial daughter cells division and endodermis differentiation. It interacts and forms transcription factor complexes with SHORTROOT (SHR), which moves from stele into adjacent cells induces the expression of SCR and regulatory components in BIRD zinc finger proteins like JACKDAW (JKD), NUTCRAKER, BLUEJAY, and MAGPIE to form SCR-SHR complexes. This leads to the activation of WOX5, CYCD6;1 or CASP1. CYC6;1 can promote formation of ground tissue from the endodermis by asymmetrical division.

Auxin is the key hormone in root development, auxin distribution patterns during the embryogenesis determines the root development. Later on, it determines the root meristem organization, regeneration, root spacing. Auxin exists as a gradient in the root tip, with highest levels in the quiescent center. Its role is to maintain the cell division and columella stem cell identity and columella shedding. It induces the expression of PLT proteins which controls the transport of auxin through PIN transporters to maintain and control elongation, cell division and differentiation and adaptation to the environmental conditions. low auxin levels induce elongation of roots, slight increase in auxin levels can result in its inhibition via TRANSPORT INHIBITOR RESPONSE1 (TIR1). Initiation of lateral roots occurs from the pericycle specific regions in the xylem poles called xylem pole pericycle (XPP). Subset of XPP cells become specified to begin lateral root initiation by asymmetrical division, and subsequently the formation of meristematic lateral root primordium. The growth of this primordium takes place to develop lateral root after emergence from the overlying cells (Motte, Vanneste, & Beeckman, 2019). Jasmonates inhibit root growth but in *coi1* mutants JA does not inhibit root growth. In *COI1* mutants the regeneration after ablation of quiescent center is delayed, JA and Aux synergistically regulate tissue regeneration. In rice the Nacl treatment leads to Ethylene biosynthesis and which increases Jasmonate biosynthesis to inhibit root growth. JA downregulate OsPLT and cell division related genes. overexpression of JAZ proteins JAZ4, JAZ8, JAZ13 results in decreased sensitivity to JA. bHLH subgroups IIIe transcription factor MYC2 positively regulates Jasmonate mediated root growth inhibition. *MYC, MYC3, MYC4* triple mutants are less sensitive to JA. MYC2 represses expression of PLT gene which regulate stem cell niche. PUB10 and BTB/POZ-MATH (BPM) ubiquitinate MYC2 and which leads its degradation via 26s proteasome this interaction acts as a control on the Jasmonate mediated inhibition of root growth. MYC2 interacts with MED25 to activate transcription of target genes. UBIQUITIN-PROTEASE (UBP) deubiquitinate MYC2 to increase its stability. *UBP12* and *UBP13* mutants are more sensitive to Jasmonates. bHLH3, bHLH13, bHLH14, and bHLH17 negatively regulate JA-mediated root growth inhibition. Their mutants show stronger inhibition of roots by Jasmonates.

EIN3 and *EIL1* are ethylene responsive transcription factors that modulate ethylene response. These transcription factors integrate ethylene and Jasmonate signaling. JAZ proteins interact with EIN3 and EIL1 and repress their activity. In presence of bioactive JA the JAZ are tagged for degradation and this repression on EIN3 and EIL1 is lost. Plants overexpressing EIN3 and EIL1 Thus JA enhances ET response in root development. JA represses expression of Brassinosteroid biosynthetic gene *DWARF4* (DWF4*)*. Arabidopsis lines with mutant *DWF4* overproducing BR showed reduced inhibition of root growth after treatment with MeJA, exogenous application of Brassinosteroids reduce the inhibitory effect of JA on roots. Both JA and BR interact mutually to optimize growth and defense responses relevant to environmental conditions. JAZ proteins interact with repressors of GA signaling DELLA proteins to prevent JAZ from inhibiting MYC2 to promote expression of JA responsive genes (Figure 2.5).

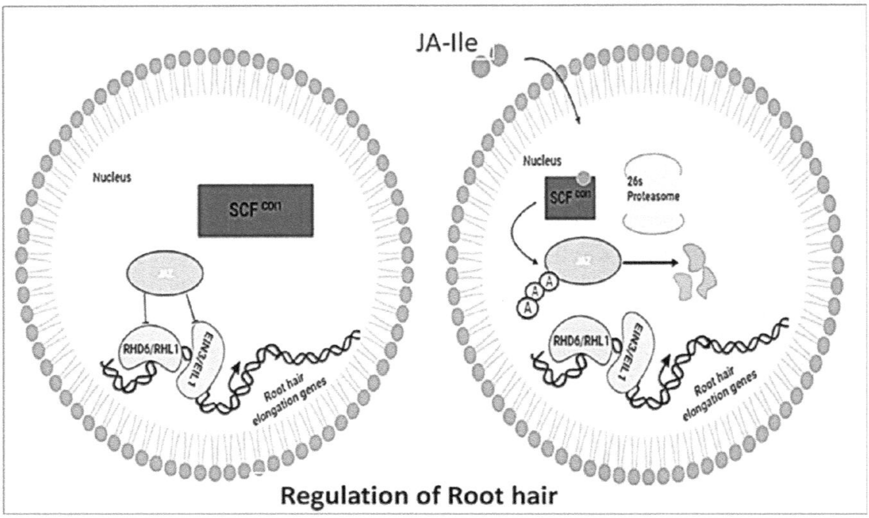

FIGURE 2.5 Depicts the suppression of RHD6/RSL1, EIN3/EIL1 by JAZ proteins. Furthermore, the activation of jasmonate signaling resulting in degradation of JAZ proteins and initiation of root hair formation.

Source: Modified figure from Han et al., (2022).

Jasmonate promotes root hair development. Jasmonates induces expression of the root hair growth bHLH transcription factors and their target genes. JAZ proteins interact repress activity of RHD6 (Root Hair Defective6 regulator of root hair development) which in important for root hair growth, presence of JA induces JAZ degradation and relieves RHD6 from this repression. Using Jasmonate exogenously induces root hair elongation. Ethylene responsive transcription factors EIN3/EIL1 interacts with RHD6/RSL1 (ROOT Hair Defective 6 like1). JAZ proteins interact with these complexes and prevents their activity. In presence of Jasmonates the JAZ proteins are degraded, integrating JA-signalling and ET signaling in root hair development (Han et al., 2022).

2.11 JASMONATES IN THIGMOMORPHOGENESIS

Thigrmomorphogenesis is a plants response to mechanical sensation by adapting or altering developmental patterns. Since plants cannot move, they must be possess the means to perceive the diverse mechanical stimuli from their environment that influence the survival, growth, and development, as well as appropriate response. Most plants show a slow and gradual response to touch stimuli except species like *Mumosa pudica* and *Dionaea muscipula* that show a rapid touch response. Thigmomorphogenesis is a broad term and encompasses different processes, for example, adaptation of tree growth to keep standing in the powerful winds, seed dispersal, to insect invasion, or herbivory.

The mechanisms behind mechanical perception is not completely understood, various genes have been identified and subsequently mutants have been developed that show weakened Thigmomorphogenesis. MS mechanosensitive channels (MSL mechanosensitive channel of small conductance -like and MCA (Mid1-complementing activity) that produce a chemical signal like (influx of ca^{2+} ions) by sensing the changes in membrane properties, as a result of a mechanical stimuli such as osmotic shock or a mechanical disturbance, Receptor like kinases (RLK) like theseus1, fer-4, glutamate receptors like GLR3.3, GLR3.6, that mediates systemic signals by generating electrical and ca^{2+} waves and eventually activating downstream signaling (Darwish et al., 2022).

JA signaling regulate Thigmomorphogenesis but this role is not exclusive to JA. Touch stimulated *Medicago truncatula* plants show changes that resemble the phenotype seen in Arabidopsis cev1 mutants which produce high levels of JA (Ellis et al., 2002; Stelmach et al., 1998). Changes such as increase in chlorophyll, biomass and stunted shoot growth (Tretner, Huth, & Hause, 2008). Coiling tendrils of *Bryona diocia* show higher levels of MeJA than the coiled ones, its application also elicits tendril response (Weiler et al., 1993) the increase in the levels of JA could be a consequence of influx of Ca^{2+} in the cytosol, which activates the phospholipase D activity in the membranes of chloroplast resulting in an accumulation of fatty acids inside chloroplast (Chehab, Eich, & Braam, 2008). Supported by the studies showing lack of touch-induced phenotypes in mutants defective in JA biosynthesis and perception like *aos, jar1*, and *coi1* for example touch treated gl-1 displayed decrease in rosette leave diameter and shorter inflorescences. conversely repetitive touch treated *aos* mutants had no effect on inflorescence length and showed no delay in flowering compared to gl-1. Similarly, in *jar1* and *coi1* mutants the repetitive touch treatment did not show any observable difference in inflorescence elongation. Mechanical stress like touch increases JA levels in shoots, even in the (OPR3-OE) transgenic lines that over express JA biosynthetic gene. OPR3-OE lines possesses a 30% higher JA levels show a strong thigmomorphogenetic response. Indicating the importance of JA signaling in Thigmomorphogenesis. Touch stimulated plants showing higher levels of JA, are primed for defense and show enhanced resistance to pathogens (Chehab et al., 2012).

JA signaling/biosynthesis mutants show defective thermomorphogenesis, the expression of mechanoreceptive genes like TCH2, TCH3 and cell wall modifying, TCH4 was induced in JA mutants. Indicating thigmorphogenesis is not exclusively regulated by JA signaling, but indeed a JA independent signaling in plants is also involved. Feronia a receptor like kinase is a negative regulator of JA dependent touch signaling and fer-4 mutants show higher levels of touch marker genes like bHLH19, CML39, JAZ7. In *fer* mutants JA signaling components were higher after touch treatment, proving the negative regulation of JA signaling by Feronia. Touch marker gene levels were elevated even in absence of touch stimulation. CAMTA (calmodulin-binding Transcriptional activator) are key factors in JA independent signaling. CAMTA1, CAMTA2, CAMTA3 bind to promoters and bring about expression of target genes like TCH2/4. With CAMTA3 being the most crucial, transgenic lines with mutated *camta3* are unable to show Thigmomorphogenesis and are completely insensitive to touch stimulation (Darwish et al., 2022).

2.12 JASMONATES AND SEED GERMINATION

Jasmonates (JAs) have been found to inhibit seed germination similarly to abscisic acid (ABA). In *Arabidopsis*, increasing 12-oxophytodienoic acid (OPDA) levels during seed development inhibits seed germination (Dave et al., 2011). Exogenous application of methyl jasmonate (MeJA) and JA inhibits seed germination in various plants, such as *Arabidopsis*, *Brassica*, and maize (Dave et al., 2011; Norastehnia, Sajedi, and Nojavan-Asghari, 2007; Wilen et al., 1991). ABA is known for suppressing seed germination and promoting seed dormancy. ABA is perceived by the regulatory component of abscisic acid receptor (RCAR)/pyrabactin resistance (PYR)/PYR1-like (PYL). These receptors bind to PP2C, forming a PYR/RCAR-PP2C complex, which releases SnRK2 from PP2C inhibition (Cutler et al., 2010; Ma et al., 2009; Miyazono et al., 2009; Nishimura et al., 2009; Santiago et al., 2009). SnRK2 auto-phosphorylates to activate itself and phosphorylate downstream transcription factors like ABI3, ABI4, and ABI5 (Kobayashi et al., 2005). These factors bind to the promoters of their target genes, such as *EM1* and *EM6* (early methionine-labeled), to suppress gene expression (Hu et al., 2019). They also converge different signaling pathways by interacting with other factors to modulate gene expression. For example, ABI3 interacts with PIL5 to induce the expression of *SOM* in imbibed *Arabidopsis* seeds by binding to its promoter (Park et al., 2011). Jasmonate signaling assists ABA in its inhibitory role in regulating seed germination, with a bidirectional cross-talk between the two pathways. Various components of ABA signaling are induced by JA. An increase in oxylipins was found in seeds during the late maturation phase in *Arabidopsis*, and exogenous OPDA (a precursor of JA) treatment increased ABI5 levels in *Arabidopsis*. OPDA is a more effective inhibitor of seed germination than MeJA and JA-Ile, and this inhibition is independent of the *COI1*-mediated pathway. It has been suggested that JAZ proteins (repressors), which are part of jasmonate signaling, inhibit the activity of downstream transcription factors like MYC2, MYC3, and MYC4 in the absence of JA. Plants overexpressing the *MYC2* transcription factor showed hypersensitivity to ABA and increased inhibition of seed germination, indicating cross-talk between JA and ABA. JAZ proteins have been shown to interact with factors like ABI3 and ABI5 to suppress the expression of downstream targets. The N-terminal of ABI3 is necessary for interacting with the middle region of JAZ proteins, which contains the ZIM domain. Overexpression of ABI3 and ABI5 simultaneously suppressed ABA-insensitive mutants of *COI1*. It has been shown that the use of exogenous MeJA on *Col-0* showed no inhibitory effect on germination; however, the combined use of MeJA and ABA resulted in a lower percentage of seed germination compared to MeJA or ABA alone, indicating that JA enhances ABA signaling to inhibit seed germination. Exogenous application of ABA showed an increase in both ABA and JA biosynthesis genes. An increase in JA levels leads to the degradation of JAZ proteins, thereby activating ABI3 and ABI5 to inhibit seed germination.

Although exogenous application of JA inhibits seed germination at non physiological concentration. Accumulation of endogenous JA have shown quite the opposite effect on seed germination. Exposure to wounding or herbivory leads of reduction of seed dormancy due to accumulation of JA-Ile. overexpression of AOS induced increase in JA-Ile during seed development reduces seed dormancy in a similar fashion. stratification of *Triticum aestivum* seeds where increase in endogenous levels JAs promoted germination after cold treatment (Xu et al., 2016) (Figure 2.6).

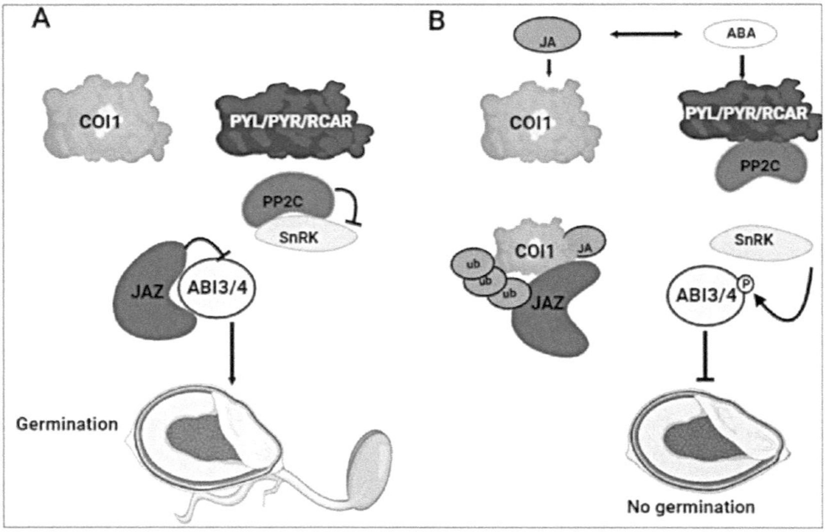

FIGURE 2.6 Depicting cross-talk between JA and ABA in regulation of seed germination. In A, JAZ proteins bind with ABI3/5 to allow germination. In B, JA signaling targets JAZ repressors to degradation and ABI3/5 are activated by ABA signaling to prevent germination.

Source: Modified figure from Varshney & Majee (2021).

Embryo development in JA biosynthesis and perception mutants is hampered, in some cases like jai1 mutant of tomato that possess mutant COI1, produces unviable seeds however wound induced OPDA can restore the seed development (Huang et al., 2017). Jasmonate (JA) negatively regulates the size of the seeds. In Arabidopsis *coi1*-2 and *coi1-8*, myc2/3/4/5 quadruple mutants had larger seeds then COL-0 consistently in *Oryza sativa OSCOI1-RNA1* mutants *con1-13* and *coi1-18* exhibited larger seeds. JAZ6 mutations could decrease the seed increased seed size in *COI1-2* mutants indicating the regulation of seed size by JA takes place in a COI dependent manner (Hu et al., 2021).

2.13 TRICHOME DEVELOPMENT

Trichomes are plant appendages found on the aerial parts of plants, such as floral organs, leaves, and stems. They differentiate from epidermal cells. While their function is not indispensable, trichomes play important roles in plant survival by acting as a physical barrier to pathogens and parasites and by providing protection from DNA-damaging ultraviolet (UV) radiation. A diverse range of trichomes exists in different plants; they can be unicellular, multicellular, glandular, non-glandular, branched, unbranched, starry, or squamiform. Glandular trichomes are involved in the synthesis and accumulation of various secondary metabolites for defense purposes. These structures form from epidermal cells in shoots. The development of trichomes from the epidermis results from a process called endoreplication or endoreduplication, in which DNA replication occurs without cell division. Endoreduplication is a major source of endopolyploidy in plants and is necessary for trichome development. More than 70 genes have been identified in trichome development, most of which are transcription factors from the MYB, basic helix-loop-helix (bHLH), and homeodomain-leucine zipper (HD-ZIP) families, playing critical roles in trichome formation. In epidermal pavement cells, *EGL3*, *GL3*, *GL1* (Glabra and Enhancer of Glabra), and *TTG1* (Transparent Testa Glabra1) form a trimeric MBW complex (MYB/bHLH/WD). This complex promotes trichome development by activating the transcription of genes like *GL2* and *TTG2* (positive regulators of trichome formation). Other genes, such as *Siamese* (SIM) and *Retinoblastoma Related1* (RBR1), downregulate CDKs, *CCS52A1*, and *CCS52A2* (Cell Cycle Switch) to initiate endoreplication. The negative regulators of trichome formation include *CAPRICE* (CPC), *Tryptychon* (TRY), Enhancer Of try and CPC 1 (ETC1), *ETC2*, and *TCL2* (trichome less). These genes migrate to neighboring cells to prevent trichome formation by establishing a repressor complex GL3/EGL3-CPC/TRY-TTG1. Mutations in these repressor genes increase trichome density, reduce trichome clustering, and decrease root hair production (Fambrini and Pugliesi, 2019).

Trichome elongation has been observed to integrate with trichome branching. Genes involved in trichome branching also participate in trichome elongation. The transcription factor *AtTCP14* regulates endoreduplication; its inhibition promotes trichome branching. *GhTCP14*, a homologue of *AtTCP14* in cotton, regulates fiber elongation. Mutant *HDG11* in *Arabidopsis* promotes trichome branching, and its homologous gene *HOX3* in cotton promotes trichome elongation (Hua et al., 2021).

2.13.1 JASMONATES IN REGULATION OF TRICHOME DEVELOPMENT

Phytohormones like Gibberellins, cytokinin, and Jasmonates act synergistically in trichome formation. JA participate in the regulation of trichome development in plants universally from solanum *lycospermum* to Gossipiyum *herbaceum.* JAZ proteins bind to regulators of trichome development GL1, EGL3, GL3 to prevent formation of MBW complex and inhibit trichome formation. In presence of JA the JAZ proteins are degraded to releasing these regulators to mediate trichome development (Pattanaik et al., 2014). JA induce defense response and trichomes are crucial in plants chemical defense towards invading organisms. Use of 50 uM of MeJA in tomato enhanced trichome elongation. Overexpression of JAZ proteins like *Gh*JAZ2 inhibited cotton fiber elongation. Conversely, overexpression *Gh*JAZ3 in Arabidopsis led to longer trichomes. *Sl*JAZ4 is the most expressed of all JAZ proteins in tomato. Tomato plants with mutant lipoxygenase gene and subsequently deficient in JA biosynthesis show abnormal development as well as reduced number of glandular trichomes (Hua et al., 2021). In jai1-1 mutants the immature fruits are completely free from trichomes with smooth appearance moreover type-VI glandular trichome (containing short stock and a four celled glandular head) were reduced in the leaves and petals (Li et al., 2004). Constitutive expression of soyabean AOS an intermediatory gene in the octa-decanoid pathway led to an increase in the trichome density on transgenic tobacco compared to the controls (Wu et al., 2008). Many transcription factors that link JA signaling with trichome development are known, e.g., *Sl*HD8 interacts with multiple JAZ proteins. *Sl*HD8 a homologue of *Gh*HOX3 in cotton. Knock down of GhHOX3 in cotton led to a reduction of trichomes by 80%. *Sl*MYC1 is a participant of JA signaling in trichome cells, interacts with SlJAZ4 and is thought to be involved in stalk and glandular cell expansion in type VI trichomes (Hua et al., 2021). use of exogenous MeJA induced expression of GL3 which is a component of the MBW complex in the development of trichomes (Yoshida et al., 2009).

2.14 JASMONATES IN FRUIT RIPENING/DEVELOPMENT

Through the process of development of fruits from flowers after fertilization, there are many changes this complex organ has to go through. Changes in fruit color, texture, aromatic substances, sugar content and developing a proper shape and size (Aubert et al., 2015). The responsibility of regulation of fruit development lies majorly on two phytohormones namely ethylene and Abscisic Acid (ABA) (Jia et al., 2011; Symons et al., 2012). Overall, in strawberry the JA levels have has been seen to decline from flowering stage to ripening (Böttcher et al., 2015; Gansser, Latza, & Berger, 1997). In *Vitis vinifera* the JA and JA-Ile levels may promote cell division and fruit set (Böttcher et al., 2015). The exogenous use of MeJA on *Fragaria chiloensis* (Chilean strawberry) results in an increase in anthocyanin levels (Concha et al., 2013). ABA and ethylene being the main regulator of fruit ripening in strawberry and tomato fruit, with its levels showing a pattern opposite to that of JA (Garrido-Bigotes, Figueroa, & Figueroa, 2018). Exogenous application of MeJA decreased ABA levels and the expression levels of FaNCED1 (which encodes an enzyme that catalyses conversion of violaxanthin to neoxanthin) in developing strawberry fruits, revealing a role of JA in negative regulation of ABA signaling (Garrido-Bigotes, Figueroa, & Figueroa, 2018).

The expression levels of JA biosynthesis genes like *FaJAR1*, *Fa13-lox*, *FaAOS*, *FaJMT*, and *FaOPR3* in strawberry showed a pattern resembling the pattern of JA levels from flowering to fruit ripening stage. The levels of JA-Ile follows a similar pattern as auxin (IAA), Brassinosteroid and Gibberellin (GA) (Garrido-Bigotes, Figueroa, & Figueroa, 2018). It is interesting to note that salicylic acid showed and undefined accumulation pattern during fruit ripening (Garrido-Bigotes, Figueroa, & Figueroa, 2018). In apple (malus domestica) JA signaling enhances ethylene production. exogenous treatment with MeJA with in fruits induced expression of a transcription factor MdMYC2. This TF binds to the G-box motif in the promoter MdACS1 and MdACO, which encode enzymes that participate in ethylene biosynthesis. It also transcriptionally up regulates ethylene biosynthesis by inducing the transcription of MdERF3 transcription factor the binds to the promoter of MdACS1. MdMYC2 interacts with MdERF2 (a repressor of MdACS1) prevent its binding at the promotor of MdACS1 and hence maintain the expression of ethylene biosynthesis genes. MdMYC2 interacts MdERF2 that would otherwise bind MdERF3 (Li et al., 2017). Additionally similar results were observed in other fruits where application of JA accelerated ripening process. In Kiwi fruit exogenous application of MeJA and ETH enhanced fruit ripening and promoted ethylene biosynthesis by regulating ACS gene thorough two NAC transcription factor genes, *AdNAC2* and *AdNAC3* (Wu et al., 2020). Jasmonates can be used to enhance the post-harvest quality and the shelf life of fruits. In papaya Fruit, treatment with MeJA vapors for 16 hours inhibited fungal decay, reduced chilling injury and firmness, levels of organic acids like citric acid, were higher in the treated fruits then the control (Gonzalez-Aguilar, Buta, & Wang, 2003). Similar results were found with use of MeJA vapors on stored mangos (González-Aguilar et al., 2000). Use of MeJA exogenously, enhanced the ripening of Japanese plum with an advancement in respiratory climacteric peak and increase in antioxidant content, carotenoids (Khan & Singh, 2007).

2.15 ACCUMULATION OF SECONDARY METABOLITES

Formation of secondary metabolites is one of the consequences of JA signaling, accumulation of endogenous JA enhances the biosynthesis of metabolites like anthocyanins, nicotine, terpenoid indole alkaloids, glucosinolates, etc. (Wasternack & Strnad, 2019). The secondary metabolites in plants are grouped into three classes: (i) terpenes/terpenoids; (ii) phenolics; and (iii) nitrogen containing compounds such as alkaloids. These metabolites are synthesized via shikimate and mevalonate pathway. Many of these metabolites possess immense pharmacological and therapeutic potential (Ali, 2021). synthesis of these metabolites in plants is spatially and temporally specific, their production large bioreactor for commercial use could help prevent the exploitation of plants species that produce them.

JA can stimulate their biosynthesis both *in-vivo* and *in-vitro* plant tissue cultures. The most well studied transcription factor in JA signaling is MYC2. It regulates flavonoid biosynthesis genes both directly and indirectly in *Arabidopsis thaliana*. In *Catharanthus roseus*, *MYC2* homologs bind to the promotors of *ORCA* transcription factors, indirectly controlling biosynthesis of Terpenoid indole alkaloids (TIA). *MYC2* regulates biosynthesis of Nicotine in *Nicotiana tabacum* by upregulating AP2/ERF TF. In *Arabidopsis*, JAZ proteins interact with various transcription factors that regulate secondary metabolite synthesis. In *Cantharanthus roseus* JA regulate the expression of DAT (deacetylvindoline 4-O-acetyltransferase) that catalyzes the formation of vindoline from deacetylcindoline by acetylation. Vindoline is an anti-cancer compound synthesized via TIA pathway. Strictosidine an intermediate in the TIA biosynthesis. STR (strictosidine synthetase) is induced by JA treatment. Other transcription factors like ORCA transcription factors regulate their biosynthesis elicited by JA. In Nicotiana JA regulate the biosynthesis of nicotine which is synthesized by coupling of pyrrolidone ring and pyridine ring after their biosynthesis. JA signaling upregulates the expression of putrescine N-methyltransferase (*PMT*) a nicotine biosynthetic gene. *Artesimisia annua* produces a lactone endoperoxide called Artemisinin (sesquiterpene). Used for treatment of malaria. Jasmonate responsive transcription factors *AaERF1* and *AaERF2* can activate the transcription of artemisinin biosynthesis genes like amorpha-4,11-diene synthase (ADS which converts farnesyl diphosphate into amorpha-4,11-diene, first committed step in biosynthesis of artemisinin) and CYP sesquiterpene oxidase (CYP71AV1). JA responsive AaWRKY1 can induce artemisinin biosynthesis by activating ADS transcription. The promotor of *ADS* gene contains cis acting W-box elements for the binding of AaWRKY1. Transient expression of this gene enhanced the expression of many artemisinin biosynthesis gene (Afrin, Huang, & Luo, (2015). In Brassicaceae the Glucosinolates are induced mechanical wounding and herbivory, they are important for the defense in Brassicaceae. Glucosinolates and Myrosinases (their hydrolytic enzymes) Normally Myrosinases and substrates are stored in different compartments and come in contact upon disruption of tissue due to mechanical damage which results in formation of toxic compounds like thiocyanates, isothiocyanates and nitriles (Winde & Wittstock, 2011). The biosynthesis of these Glucosinolate regulated by MYB28, MYB29 and MYB76 (for aliphatic Glucosinolates) and MYB34, MYB51 MYB122 (for indole Glucosinolates) these MYB TFs are themselves regulated by MYC2 and other MYC transcription factors. *myc2 myc3*, *myc4* triple mutants show reduced Glucosinolate levels. MYC2 also directly regulates biosynthesis of Glucosinolates by binding to the gene involved in its biosynthesis (Chen et al., 2019).

In vitro tissue culture methods can be an alternative for the production of secondary metabolites that cannot be chemically synthesized due to complex chemical structures. The production of metabolites could be further increased by the use of elicitors. Elicitors are substances that are used to enhance the production of biomass and/or metabolites. They can be of biotic or abiotic origin, for example, plant cell wall fragments, some chemical compounds, growth hormones, or some physical treatments like pH, osmotic stress, temperature, and ionic stress (e.g., Cd, Cu, Ag). Jasmonates have been used as elicitors in *in vitro* plant tissue cultures. Methyl jasmonate (MeJA) has been found to induce the expression of phenyl ammonia lyase (PAL), which is required for the production of lignins, anthocyanins, and coumarins, as well as chalcone isomerase (CHI). MeJA downregulates catalase (CAT) activity (CAT enzyme catalyzes reactive oxygen species (ROS)). Increased ROS levels act as secondary messengers and induce the production of secondary metabolite synthesis (Nabi, Singh, & Saffeullah, 2021). In *Allium jesdianum*, the use of MeJA was more effective than putrescine in increasing secondary metabolite production. However, a higher concentration of MeJA reduced the weight of fresh callus, opposite to putrescine, which enhanced the growth of the callus (Yazdanian et al., 2022). *Castilleja tenuiflora* Benth., a medicinal plant, showed an increase in total phenolic compounds upon exposure to different concentrations of MeJA and increased the expression of genes such as 1-deoxy-D-xylulose-5-phosphate synthase (DXS), geraniol 10-hydroxylase (G10H), PAL, and chalcone synthase (CHS), without affecting the growth of plants. The enhancement in the production of secondary metabolites (SMs) was accompanied by hydrogen peroxide (H_2O_2) production. From a range of concentrations (50–150 μM), the highest production was seen at 100 μM of MeJA (Rubio-Rodríguez et al., 2021). In another study, elicitation by MeJA was more effective than salicylic acid (SA) in the production of cucurbitacins, using mevalonic acid and squalene as precursors in cell suspension cultures and *in vitro* plant cultures of *Citrullus colocynthis* (L.) (Dasari et al., 2020).

KEYWORDS

- **abiotic stress**
- **biosynthesis gene**
- **jasmonic acid**
- **secondary metabolites**
- **signaling**
- **trichome development**

REFERENCES

Acosta, I. F., & Przybyl, M. (2019). Jasmonate signaling during *Arabidopsis* stamen maturation. *Plant and Cell Physiology, 60*(12), 2648–2659.

Afrin, S., Huang, J. J., & Luo, Z. Y. (2015). JA-mediated transcriptional regulation of secondary metabolism in medicinal plants. *Science Bulletin, 60*(12), 1062–1072.

Ali, B. (2021). Practical applications of jasmonates in the biosynthesis and accumulation of secondary metabolites in plants. *Biocatalysis and Agricultural Biotechnology, 38*, 102205.

Aubert, Y., et al. (2015). CYP94-mediated jasmonoyl-isoleucine hormone oxidation shapes jasmonate profiles and attenuates defense responses to *Botrytis cinerea* infection. *Journal of Experimental Botany, 66*(13), 3879–3892.

Böttcher, C., et al. (2015). Jasmonic acid-isoleucine formation in grapevine (*Vitis vinifera* L.) by two enzymes with distinct transcription profiles. *Journal of Integrative Plant Biology, 57*(7), 618–627.

Brioudes, F., et al. (2009). Jasmonate controls late development stages of petal growth in *Arabidopsis thaliana. The Plant Journal, 60*(6), 1070–1080.

Browse, J., & Wallis, J. G. (2019). *Arabidopsis* flowers unlocked the mechanism of jasmonate signaling. *Plants, 8*(8), 285.

Cai, Y., et al. (2020). The HD-ZIP IV transcription factor GL2-LIKE regulates male flowering time and fertility in cucumber. *Journal of Experimental Botany, 71*(18), 5425–5437.

Caldelari, D., et al. (2011). *Arabidopsis* lox3 lox4 double mutants are male sterile and defective in global proliferative arrest. *Plant Molecular Biology, 75*(1), 25–33.

Chehab, E. W., Eich, E., & Braam, J. (2008). Thigmomorphogenesis: A complex plant response to mechano-stimulation. *Journal of Experimental Botany, 60*(1), 43–56.

Chehab, E. W., et al. (2012). *Arabidopsis* touch-induced morphogenesis is jasmonate-mediated and protects against pests. *Current Biology, 22*(8), 701–706.

Chen, X., et al. (2019). Plant specialized metabolism regulated by jasmonate signaling. *Plant and Cell Physiology, 60*(12), 2638–2647.

Chung, H. S., & Howe, G. A. (2009). A critical role for the TIFY motif in repression of jasmonate signaling by a stabilized splice variant of the JASMONATE ZIM-domain protein JAZ10 in *Arabidopsis. The Plant Cell, 21*(1), 131–145.

Concha, C. M., et al. (2013). Methyl jasmonate treatment induces changes in fruit ripening by modifying the expression of several ripening genes in *Fragaria chiloensis* fruit. *Plant Physiology and Biochemistry, 70*, 433–444.

Cutler, S. R., et al. (2010). Abscisic acid: Emergence of a core signaling network. *Annual Review of Plant Biology, 61.*

Darwish, E., et al. (2022). Touch signaling and thigmomorphogenesis are regulated by complementary CAMTA3- and JA-dependent pathways. *Science Advances, 8*(20), eabm2091.

Dasari, R., et al. (2020). Enhancement of production of pharmaceutically important anti-cancerous compound; cucurbitacin E via elicitation and precursor feeding of *in vitro* culture of *Citrullus colocynthis* (L.) Schard. *Vegetos, 33*(2), 323–334.

Dave, A., et al. (2011). 12-Oxo-phytodienoic acid accumulation during seed development represses seed germination in *Arabidopsis. The Plant Cell, 23*(2), 583–599.

Ding, F., et al. (2022). SlMYC2 mediates jasmonate-induced tomato leaf senescence by promoting chlorophyll degradation and repressing carbon fixation. *Plant Physiology and Biochemistry, 180*, 27–34.

Dobritzsch, S., et al. (2015). Dissection of jasmonate functions in tomato stamen development by transcriptome and metabolome analyses. *BMC Biology, 13*(1), 1–18.

Ellis, C., et al. (2002). The *Arabidopsis* mutant cev1 links cell wall signaling to jasmonate and ethylene responses. *The Plant Cell, 14*(7), 1557–1566.

Fambrini, M., & Pugliesi, C. (2019). The dynamic genetic-hormonal regulatory network controlling the trichome development in leaves. *Plants, 8*(8), 253.

Feys, B. J., et al. (1994). *Arabidopsis* mutants selected for resistance to the phytotoxin coronatine are male sterile, insensitive to methyl jasmonate, and resistant to a bacterial pathogen. *The Plant Cell, 6*(5), 751–759.

Fu, W., et al. (2015). Acyl-CoA N-acyltransferase influences fertility by regulating lipid metabolism and jasmonic acid biogenesis in cotton. *Scientific Reports, 5*(1), 1–17.

Gansser, D., Latza, S., & Berger, R. G. (1997). Methyl jasmonates in developing strawberry fruit (*Fragaria ananassa Duch. Cv. Kent*). *Journal of Agricultural and Food Chemistry, 45*(7), 2477–2480.

Garrido-Bigotes, A., Figueroa, P. M., & Figueroa, C. R. (2018). Jasmonate metabolism and its relationship with abscisic acid during strawberry fruit development and ripening. *Journal of Plant Growth Regulation, 37*(1), 101–113.

Goetz, S., et al. (2012). Role of cis-12-oxo-phytodienoic acid in tomato embryo development. *Plant Physiology, 158*(4), 1715–1727.

Gonzalez-Aguilar, G., Buta, J. G., & Wang, C. (2003). Methyl jasmonate and modified atmosphere packaging (MAP) reduce decay and maintain postharvest quality of papaya 'Sunrise.' *Postharvest Biology and Technology, 28*(3), 361–370.

Gonzãlez-Aguilar, G., et al. (2000). Methyl jasmonate reduces chilling injury and maintains postharvest quality of mango fruit. *Journal of Agricultural and Food Chemistry, 48*(2), 515–519.

Guo, Q., et al. (2018). JAZ repressors of metabolic defense promote growth and reproductive fitness in *Arabidopsis*. *Proceedings of the National Academy of Sciences, 115*(45), E10768–E10777.

Han, X., et al. (2022). Jasmonate-regulated root growth inhibition and root hair elongation. *Journal of Experimental Botany, 74(4), 1176–1185.*

Heil, M. (2011). Nectar: Generation, regulation and ecological functions. *Trends in Plant Science, 16*(4), 191–200.

Hu, S., et al. (2021). Control of seed size by jasmonate. *Science China Life Sciences, 64*(8), 1215–1226.

Hu, Y., et al. (2017). Jasmonate regulates leaf senescence and tolerance to cold stress: Crosstalk with other phytohormones. *Journal of Experimental Botany, 68*(6), 1361–1369.

Hu, Y., et al. (2019). The transcription factor inducer of cbf expression1 interacts with abscisic acid insensitive5 and Della proteins to fine-tune abscisic acid signaling during seed germination in *Arabidopsis*. *The Plant Cell, 31*(7), 1520–1538.

Hua, B., et al. (2021). Homeodomain protein8 mediates jasmonate-triggered trichome elongation in tomato. *New Phytologist, 230*(3), 1063–1077.

Huang, H., et al. (2014). Amino acid substitutions of GLY98, LEU245, and GLU543 in COI1 distinctively affect jasmonate-regulated male fertility in *Arabidopsis*. *Science China Life Sciences, 57*(1), 145–154.

Huang, H., et al. (2017). *Arabidopsis* MYB24 regulates jasmonate-mediated stamen development. *Frontiers in Plant Science, 8*, 1525.

Huang, H., et al. (2017). Jasmonate action in plant growth and development. *Journal of Experimental Botany, 68*(6), 1349–1359.

Irish, V. (2017). The ABC model of floral development. *Current Biology, 27*(17), R887–R890.

Jia, H. F., et al. (2011). Abscisic acid plays an important role in the regulation of strawberry fruit ripening. *Plant Physiology, 157*(1), 188–199.

Jiang, Y., et al. (2014). *Arabidopsis* WRKY57 functions as a node of convergence for jasmonic acid– and auxin-mediated signaling in jasmonic acid–Induced leaf senescence. *The Plant Cell, 26*(1), 230–245.

Khan, A. S., & Singh, Z. (2007). Methyl jasmonate promotes fruit ripening and improves fruit quality in Japanese plum. *The Journal of Horticultural Science and Biotechnology, 82*(5), 695–706.

Kinoshita, A., & Richter, R. (2020). Genetic and molecular basis of floral induction in *Arabidopsis thaliana. Journal of Experimental Botany, 71*(9), 2490–2504.

Kobayashi, Y., et al. (2005). Abscisic acid-activated SNRK2 protein kinases function in the gene-regulation pathway of ABA signal transduction by phosphorylating ABA response element-binding factors. *The Plant Journal, 44*(6), 939–949.

Li, C., et al. (2003). The tomato suppressor of prosystemin-mediated responses2 gene encodes a fatty acid desaturase required for the biosynthesis of jasmonic acid and the production of a systemic wound signal for defense gene expression. *The Plant Cell, 15*(7), 1646–1661.

Li, L., et al. (2004). The tomato homolog of coronatine-insensitivel is required for the maternal control of seed maturation, jasmonate-signaled defense responses, and glandular trichome development. *The Plant Cell, 16*(1), 126–143.

Li, Q., et al. (2017). Transporter-mediated nuclear entry of jasmonoyl-isoleucine is essential for jasmonate signaling. *Molecular Plant, 10*(5), 695–708.

Li, R., et al. (2018). Jasmonate signaling makes flowers attractive to pollinators and repellent to florivores in nature. *Journal of Integrative Plant Biology, 60*(3), 190–194.

Li, T., et al. (2017). The jasmonate-activated transcription factor MdMYC2 regulates ethylene response factor and ethylene biosynthetic genes to promote ethylene biosynthesis during apple fruit ripening. *The Plant Cell, 29*(6), 1316–1334.

Ma, Y., et al. (2009). Regulators of PP2C phosphatase activity function as abscisic acid sensors. *Science, 324*(5930), 1064–1068.

Mandaokar, A., & Browse, J. (2009). MYB108 acts together with MYB24 to regulate jasmonate-mediated stamen maturation in *Arabidopsis*. *Plant Physiology, 149*(2), 851–862.

Miyazono, K., et al. (2009). Structural basis of abscisic acid signaling. *Nature, 462*(7273), 609–614.

Motte, H., Vanneste, S., & Beeckman, T. (2019). Molecular and environmental regulation of root development. *Annual Review of Plant Biology, 70*, 465–488.

Murofushi, N., et al. (1999). 8.02 – Plant hormones. In S. D. Barton, K. Nakanishi, & O. Meth-Cohn (Eds.), *Comprehensive Natural Products Chemistry* (pp. 19–136). Oxford: Pergamon.

Nabi, N., Singh, S., & Saffeullah, P. (2021). Responses of *in vitro* cell cultures to elicitation: Regulatory role of jasmonic acid and methyl jasmonate: A review. *In Vitro Cellular & Developmental Biology-Plant, 57*, 341–355.

Nicolson, S. W., Nepi, M., & Pacini, E. (2007). *Nectaries and Nectar* (Vol. 4). Springer.

Nishimura, N., et al. (2009). Structural mechanism of abscisic acid binding and signaling by dimeric PYR1. *Science, 326*(5958), 1373–1379.

Niwa, T., et al. (2018). Jasmonic acid facilitates flower opening and floral organ development through the upregulated expression of SlMYB21 transcription factor in tomato. *Bioscience, Biotechnology, and Biochemistry, 82*(2), 292–303.

Norastehnia, A., Sajedi, R., & Nojavan-Asghari, M. (2007). Inhibitory effects of methyl jasmonate on seed germination in maize (*Zea mays*): Effect on α-amylase activity and ethylene production. *Gen. Appl. Plant Physiol, 33*(1, 2), 13–23.

Park, J., et al. (2011). ABI3 and PIL5 collaboratively activate the expression of *SOMNUS* by directly binding to its promoter in imbibed *Arabidopsis* seeds. *The Plant Cell, 23*(4), 1404–1415.

Pattanaik, S., et al. (2014). An overview of the gene regulatory network controlling trichome development in the model plant, *Arabidopsis*. *Frontiers in Plant Science, 5*.

Qi, T., et al. (2015). Regulation of jasmonate-mediated stamen development and seed production by a bHLH-MYB complex in *Arabidopsis*. *The Plant Cell, 27*(6), 1620–1633.

Radhika, V., et al. (2010). Regulation of extrafloral nectar secretion by jasmonates in lima bean is light dependent. *Proceedings of the National Academy of Sciences, 107*(40), 17228–17233.

Radhika, V., et al. (2010). The role of jasmonates in floral nectar secretion. *PLoS One, 5*(2), e9265.

Reeves, P. H., et al. (2012). A regulatory network for coordinated flower maturation. *PLoS Genetics, 8*(2), e1002506.

Rubio-Rodríguez, E., et al. (2021). Secondary metabolite production and related biosynthetic genes expression in response to methyl jasmonate in *Castilleja tenuiflora* Benth. *in vitro* plants. *Plant Cell, Tissue and Organ Culture (PCTOC), 144*(3), 519–532.

Rudall, P. J. (2020). *Anatomy of Flowering Plants: An Introduction to Plant Structure and Development*. Cambridge University Press.

Santiago, J., et al. (2009). The abscisic acid receptor PYR1 in complex with abscisic acid. *Nature, 462*(7273), 665–668.

Schmitt, A. J., et al. (2018). The Octadecanoid Pathway, but not COI1, is required for nectar secretion in *Arabidopsis thaliana*. *Frontiers in Plant Science, 9*, 1060.

Song, S., et al. (2011). The jasmonate-ZIM domain proteins interact with the R2R3-MYB transcription factors MYB21 and MYB24 to affect jasmonate-regulated stamen development in *Arabidopsis*. *The Plant Cell, 23*(3), 1000–1013.

Stelmach, B. A., et al. (1998). Quantitation of the octadecanoid 12-oxo-phytodienoic acid, a signaling compound in plant mechanotransduction. *Phytochemistry, 47*(4), 539–546.

Sun, B., et al. (2019). Jasmonate-inducible R2R3-MYB transcription factor regulates capsaicinoid biosynthesis and stamen development in Capsicum. *Journal of Agricultural and Food Chemistry, 67*(39), 10891–10903.

Symons, G., et al. (2012). Hormonal changes during non-climacteric ripening in strawberry. *Journal of Experimental Botany, 63*(13), 4741–4750.

Tang, B., et al. (2022). SlJAZ10 and SlJAZ11 mediate dark-induced leaf senescence and regeneration. *PLoS Genetics, 18*(7), e1010285.

Thines, B., et al. (2007). JAZ repressor proteins are targets of the *SCFCOI1* complex during jasmonate signaling. *Nature, 448*(7154), 661–665.

Tretner, C., Huth, U., & Hause, B. (2008). Mechanostimulation of *Medicago truncatula* leads to enhanced levels of jasmonic acid. *Journal of Experimental Botany, 59*(10), 2847–2856.

Varshney, V., & Majee, M. (2021). JA shakes hands with ABA to delay seed germination. *Trends in Plant Science, 26*(8), 764–766.

Wang, W., et al. (2014). The F-box protein COI1 functions upstream of MYB305 to regulate primary carbohydrate metabolism in tobacco (*Nicotiana tabacum* L. cv. TN90). *Journal of Experimental Botany, 65*(8), 2147–2160.

Wasternack, C., & Strnad, M. (2019). Jasmonates are signals in the biosynthesis of secondary metabolites—Pathways, transcription factors and applied aspects—A brief review. *New Biotechnology, 48*, 1–11.

Weiler, E. W., et al. (1993). Evidence for the involvement of jasmonates and their octadecanoid precursors in the tendril coiling response of *Bryonia dioica*. *Phytochemistry, 32*(3), 591–600.

Wilen, R. W., et al. (1991). Effects of jasmonic acid on embryo-specific processes in *Brassica* and *Linum* oilseeds. *Plant Physiology, 95*(2), 399–405.

Winde, I., & Wittstock, U. (2011). Insect herbivore counteradaptations to the plant glucosinolate-myrosinase system. *Phytochemistry, 72*(13), 1566–1575.

Woo, H. R., et al. (2019). Leaf senescence: Systems and dynamics aspects. *Annual Review of Plant Biology, 70*, 347–376.

Wu, J., et al. (2008). Constitutive overexpression of AOS-like gene from soybean enhanced tolerance to insect attack in transgenic tobacco. *Biotechnology Letters, 30*(9), 1693–1698.

Wu, Y., et al. (2020). Methyl jasmonate enhances ethylene synthesis in kiwifruit by inducing NAC genes that activate *ACS1*. *Journal of Agricultural and Food Chemistry, 68*(10), 3267–3276.

Xie, D. X., et al. (1998). COI1: An *Arabidopsis* gene required for jasmonate-regulated defense and fertility. *Science, 280*(5366), 1091–1094.

Xu, Q., et al. (2016). A role for jasmonates in the release of dormancy by cold stratification in wheat. *Journal of Experimental Botany, 67*(11), 3497–3508.

Yazdanian, E., et al. (2022). Elicitation effects on some secondary metabolites and antioxidant activity in callus cultures of *Allium jesdianum* Boiss. & Buhse.: Methyl jasmonate and putrescine. *Applied Biochemistry and Biotechnology, 194*(2), 601–619.

Yoshida, Y., et al. (2009). Jasmonic acid control of glabra3 links inducible defense and trichome patterning in *Arabidopsis*.

Zhao, L., et al. (2022). The emerging role of jasmonate in the control of flowering time. *Journal of Experimental Botany, 73*(1), 11–21.

CHAPTER 3

Crosstalk of Jasmonates with Secondary Metabolites, Antioxidants, and Secondary Messengers

YASHODHARA GOSWAMI[1] and ISHANI CHAKRABARTTY[2]

[1]*Department of Biomedical Genetics, School of Biosciences and Technology, Institute of Technology, Vellore Campus, Katpadi Road, Vellore, Tamil Nadu, India*

[2]*Learning and Development Solutions, Medical Affairs, Indegene Pvt. Ltd., Manyata Tech Park, Nagarwara, Bangalore, Karnataka, India*

ABSTRACT

Jasmonates can be defined as a class of oxylipins which are produced to protect plant tissues from pathogenic attackers, herbivores as well as mechanical/abiotic stresses like drought, salt, temperature, and others. These molecules are involved in signal transduction which can manipulate genome-wide-changes; they also play some important roles in fertility. Jasmonates and their combined forms with other molecules shave been studied to understand the signaling machinery. Some complex molecules are biosynthesized to maintain an equilibrium between growth-development-yield of the crops and homeostasis to the environmental stimuli. Chemically variable, such compounds are referred to as secondary molecules which act as bridges between primary metabolite production and their signaling pathways. Some of the secondary metabolites, both enzymatic and non-enzymatic, are defense players against oxidative stresses of nature and anthropogenic-activity origin. A number of antioxidants utilize

Biotechnological Applications of Jasmonates: A Group of Less Well-Known Plant Growth Regulators.
Tariq Ahmad Dar, Bilal Ahmad, and Khalid Rehman Hakeem (Eds.)

redox reactions to function against environmental challenges for the safety of the plant body/tissues.

Since jasmonates and secondary metabolites, both act in response to biotic-abiotic challenges, they share signals at several times. Their interactions, positive and negative regulatory actions on each other can be determined as cross-talks, which are mediated by numerous responsive genes and transcriptional factor proteins. This chapter aims at the study of mechanisms of JA and other secondary metabolites synthesis-modulations, their roles in plant defense system and interrelations/cross-talk of these passages at genetic and molecular level. Secondary messengers are the carrier molecules of the signaling pathways which play the major role in the cross-talks.

3.1 INTRODUCTION

Plants are the eukaryotic organisms containing a green pigment called chlorophyll, using which they perform photosynthesis; they are the major producers on earth which convey energy of sunlight to other organisms on earth, directly or indirectly, through the food they produce. Though they are present in different forms, the basic tissues are similar; modifications and specification of tissues happen so that they can adapt to their surrounding environment. But, numerous factors like pathogens, abiotic-mechanical stress and other environmental components can damage their tissues; A plant's growth and development, all the physiological processes, metabolic activities are regulated by some intrinsic chemicals, in both positive and adverse manner. These are known as plant hormones. A few of these hormones are also responsible for the signaling and feedback mechanisms of plant tissue injury by several threatening agents. The alteration in various life processes in response to the tissue wound using complex signal transduction pathways determine the health of the plant. One of those defense or wound hormones of plant is – Jasmonate (JA). According to the studies, JA responds to systemic damages through both autonomous and non-autonomous passageways, and its de-novo synthesis influences the wound-responses as well (Koo and Howe, 2009).

Plants produce a huge number of complex substances to utilize in response to stressful conditions as well as for the fulfilment of physiological functions. These substances have industrial and clinical applications in humankind too. They are referred to as secondary metabolites. At earlier times, secondary metabolites were thought to be substances mediating signals in response to environmental triggers, opposing to the primary plant growth and development processes. Now they are considered as metabolic networks along with the primary metabolites. These metabolites are derivatives of primary compounds of plant body, which can be engineered and manipulated at genetic and biochemical level to make a sustainable balance between biomass yield and adaptation to environmental changes. Based on chemical characteristics, they are classified into different types which are as alkaloids, phenolic compounds, sulfur-based compounds and terpenoids, and their production processes, effects are diverse under various stressful conditions and stimulations (Pérez-Llorca et al., 2023); some of them show antimicrobial, antioxidant activities. Some of the crop plants undergo genetic modifications by DNA recombination technology to enhance the synthesis of such bioactive compounds. These secondary metabolites are present in less amount inside the plant tissues; therefore, to increase their production, induction by environmental stresses is needed (Guerriero et al., 2018).

Cross-talk is the process of influencing one pathway by the components of another signal transduction pathway; it can be defined as the sharing of the same signal by two or multiple passages. Several molecules called secondary messengers are the connectors of these signaling pathways. This chapter focuses firstly on brief explanations about the secondary metabolites and their functions, from both biomolecular and genetic aspects; later it is going to describe about the interactive regulations and cross-talks between jasmonates (JA) and other metabolites. The importance of antioxidant properties as well as the cross-talk between antioxidants and JA mediated pathways will be discussed further (Larrieu and Vernoux, 2016).

3.2 SECONDARY METABOLITES AND THEIR CROSS-TALK/ REGULATORY INTERACTIONS WITH JA

3.2.1 BASIC MECHANISM OF JA BIOSYNTHESIS AND INTERACTION WITH THE SECONDARY METABOLITES

The defense hormone, Jasmonate, indulges itself to regulate the secondary metabolite and physical barrier production as well as change in vegetative growth rate, to give protection against infections and injuries. The activity of JA directly related to the gene expression pattern. Under stressful conditions, Jasmonate synthesis is activated and increased level of the hormone stimulates the repressor protein JAZ, which blocks the transcription factor for JA-responsive genes in unstressed tissues, to interact with COI1, which in turn degrades JAZ via a specific pathway. Newly developed techniques have demonstrated that not only bioactive JA, but JA conjugated with amino acid, named Jasmonoyl-L-isoleucine is the major activator of JAZ-COI1 interaction signaling and JA-dependent physiology. The amount of JA and JA-Ile in cells depend on developmental stage, environmental state, mechanical injury of the plant and wounding sends signal for accumulation of other JA-amino acid conjugates too, along with JA-Ile which are involved in wound-response. Attenuation mechanism is also taken up by plant cells for JA-mediated injury responses; a negative feedback control is involved too.

In reproductive tissues, lipases induce the release of substrates for de novo synthesis of jasmonates. Wounding stimulates this process in undamaged parts of the plant with the help of some chemical elicitors. Plant-derived signals like electrical or liquid-based signals trigger de novo JA synthesis in response to wounding or cellular instinctive stresses; the disarrangement of cell-wall integrity, sensors related to which can detect those physiological signals, in turn induces the JA synthesis mechanisms. Another cascade pathway relates MAP kinase to the activation of JA synthesis (Koo and Howe, 2009).

3.2.2 SECONDARY METABOLITES AND THEIR BIOSYNTHESIS/ SIGNALING TRANSDUCTION REGULATIONS

Secondary metabolites are very important biochemicals for maintaining plants' health since they help in adapting in stressful conditions; their synthesis is influenced by several environmental factors like humidity, heat, light, concentration of nutrients and ions, gases, etc. Various signal molecules, phytohormones and related biomolecules can cross-talk with these secondary metabolites and adversely affect their production as well as activity; the cell wall integrity is involved in the checking of physicochemical signaling and cross-talk happens between cell wall and external environment, with the involvement of membrane receptors. During the condition of abiotic stress, the cell wall responds through increased plasticity or hardening of it (Guerriero et al., 2018).

Some of the precursors of secondary metabolites are originally macromolecular in nature. These secondary metabolites may grant safety against environmental stress, pathogenic and herbivorous attacks, high radiation exposures, physical and mechanical wounding, and play important role in plant growth-development, whose physiological conditions often decide the amount of their availability and production.

Exposure to threatening conditions like drought, salt, and other nutrient stresses, causes the plants to produce and accumulate more levels of secondary metabolites such as, phenylpropanoids. Reactions like osmosis, dehydration in cells and tissues are the resultant of abiotic stresses. Certain gene expressions can be highlighted in response to environmental major stresses like salinity in soil. A few of the secondary metabolites act as antioxidants and formation of phenyl amides, polyamines in plants like tobacco, bean under abiotic stresses depicts the antioxidant activity of those metabolites (Ramakrishna and Ravishankar, 2011).

Jasmonates' exogenous induction may increase the gene expression for drought-responsive metabolites; *ABF1* and *ABF4* regulate an enzyme called trehalose-6-phosphate phosphatase which in turn controls stomatal closure. JA and MeJ stimulate biosynthesis of secondary metabolites like alkaloids, terpenoids, phenolic phytoalexins too. Antioxidant enzymes, namely gamma-glutamyl cysteine synthetase, DHAR, L-galactono-1,4-lactone dehydrogenase, etc., are also induced by jasmonates. With the coordination between reprogramming of the plant metabolic activities and opposing of stress production, the importance of cross-talk between PYL and JAZ-MYC2 can be understood. Under drought conditions, JA cross-talk with ABA is very much crucial to such stress-responses with the involvement of ascorbate-glutathione cycle. On the other hand, jasmonates cross-talk with CK is vital for drought-tolerance control. JA signal is beneficial, along with SA concentrations. MeJ enhances antioxidant activities, slow up leave senescence, recovery of biosynthetic rate. Jasmonates also boost up proline levels, ROS rummaging and decrease LOX activity. In rice plants, over and under-expression of specific jasmonates related genes affect the drought-tolerance (Jogawat et al., 2021). The jasmonates are also involved in regulation of energy metabolites like NAD, NADP. Adenosine, etc. Under salt-stresses, adenosine-eliminating enzymes are activated. JA diminishing plants show depletion in NAD, NADP production too, in turn altering respiratory ATP synthesis, ROS regulation and reductive biosynthesis. JA is a regulator of phytological responses in low temperature (Savchenko et al., 2019).

In case of sulfur deficiency, though jasmonates cannot directly stimulate the expression of GFP, it is observed that in that deficit condition,

JA-synthesis genes are upregulated; MeJ is a regulator of S-assimilation enzymes like APR and SAT. Under abiotic stress, Jasmonates increase the secondary metabolites. Under the condition of arsenic stress, MeJ induces the higher expression of phenylpropanoid pathway associated genes. An enzyme called phenylalanine ammonia lyase is the key component of phenylpropanoid pathway. In *Fagopyrum tataricum*, it has been noticed that JA-responsive MYB factors suppress that key enzyme and thus rutin biosynthesis gets inhibited; but, *FtJAZ* and *FtMYB* degeneration stimulate transcription and rutin biosynthesis. MeJ trigger the expression of terpene synthase genes; Other secondary metabolites such as, anthraquinone, lignin, saponin are biosynthetically upregulated with the help of methyl jasmonates applications. The same compound can upregulate the biosynthesis of sterol precursor – squalene (Iqbal et al., 2013).

Terpenoids are the largest heterogeneous chemicals which include chlorophylls, carotenoids, tocochromanols, phytohormones like gibberellin, etc. Under different abiotic stresses like drought, cold, salt, a number of terpene synthases genes' sensitivity and expression are increased while some of these genes are downregulated too. In seedlings of *Ricinus communis* and *Camellia sinensis*, *46 RcTS*-Like genes and 80 TS-like genes have been found, respectively, out of which some are highly coding and responsive to abiotic stresses. Most of the phenolic compounds are flavonoids which act as antioxidants too. Plants facing heat and cold stress generally synthesize more flavonoids, anthocyanins, phenolic acids through the maintained expression of genes of biosynthetic pathway. Higher production of such phenolic compounds is a reason of betterment of crop quality too. Specialized secondary metabolites called glucosinolates show fluctuating level under light and nutrient stress as well as other abiotic stresses, in the family Brassicaceae. They find regulatory applications in stomatal movements and Aux/IAA repressors IAA6, IAA5, IAA19 are the regulators of accumulation of this secondary metabolite. CYP79F1, CYP79F2 of aliphatic glucosinolate pathway, HOS1 are the regulatory genes and ammonium is the stimulator of GSL metabolism. Elicitation is a technique to improvise the production of secondary metabolites, using signaling molecules. Jasmonates and specifically its derivative MeJ cross-talk to induce gene transcription and biosynthesis of a number of secondary metabolites. JA elicits through targeting on specific binding to cis-acting elements at the promoter region of secondary metabolite associated genes of interest. Upregulation of signaling pathways by acting on alkaloid producing genes, phenylpropanoid pathway genes, flavonoid biosynthesis in hairy-root cultures can be achieved by methyl JA elicitation. ERF family transcription factors like *ERF210*, *ERF199*, and a few more are major components of nicotine biosynthesis and they are also stimulated by Methyl JA. Derivatives of jasmonates like Methyl jasmonate, and others may elicit the production of a number of secondary metabolites. They induce various signaling pathways, showing morphological effects. Accumulation of secondary metabolites like shikonin derivatives in *Onosma paniculatum*, endogenous IAA in plants, biogenic amines are induced by MeJ content. In plants like wild carrots, MeJ and salicylic acid generate anthocyanin production.

Elicitors activate the secondary signaling pathways; jasmonates are the major signal transduction regulators which maintain the balance of biosynthesis and response of secondary metabolites under challenging conditions. TFs are tools in modulation of biosynthetic mechanisms of secondary metabolites. CrMYC2 is activated that is involved in JA-induced degradation of JAZ. Nicotine accumulation is supported by the COI1-JAZ mechanism. In medicinal plants, joint contribution of AP2/ERF and bHLH Transcriptional factors control JA-induced elicitation of nicotine biosynthesis (Sohn et al., 2022).

3.2.3 JA/SA CROSS-TALK IN PRESENCE OF BIOTIC AND ABIOTIC STRESSES

Plants are always exposed to several types of pathogens and abiotic factors which can deteriorate the physiological condition or health of the plants; different association of attackers have to be treated in variable manner to protect the plant from their ill actions (Figure 3.1). Still, in a broader perspective, these defense mechanisms are divided into two classes – against biotrophic pathogens, and against herbivores and necrotrophs. Plants cannot locomote and therefore get attacked by numerous pathogens. Biotrophs are the pathogens which invade the epidermal tissues of plant body and divide using nutrients from host body. Salicylic acid is an essential molecule that alters host-pathogen susceptibility and confers defense against biotrophic pathogens using signal transduction pathways. Again, necrotrophs can kill the hosts by releasing toxic chemicals and feed on the remaining parts. Jasmonates/ethylene mediated signaling pathway gives protection against the later type of pathogens. Both the signaling pathways have a cross-talk strategy at different levels, including protein modification, genomic expression and plant growth-regulation mechanisms associated with phytohormones.

The inducible defenses include trichome structures, toxins like alkaloids production, enzyme inhibitors, secondary defenses like plant volatile releases, etc. These two defense pathways are majorly regulated by jasmonates and salicylic acid signaling and their cross-talk eventually results into reciprocal antagonism. The genetic basis of this cross-talk induced antagonism is very complex. This study has been done using a model plant called *Arabidopsis thaliana*. ET plays a major role in this cross-talk since it modifies NPR1 protein effect which is essential in SA-inducing pathway and the plants' JA-responsive gene expression prioritization in response to diverse assailants. NPR1 can prevent the SA signaling from repressing JA-induced defense mechanism; hence, can be recognized as a negative regulator of this cross-talk. Several studies have tried to explain the genetic and biochemical basis of this cross-talk by eliciting one pathway and observing the change in other signaling with the use of exogenous treatment of JA and SA, either individually or in different combinations (Figure 3.1).

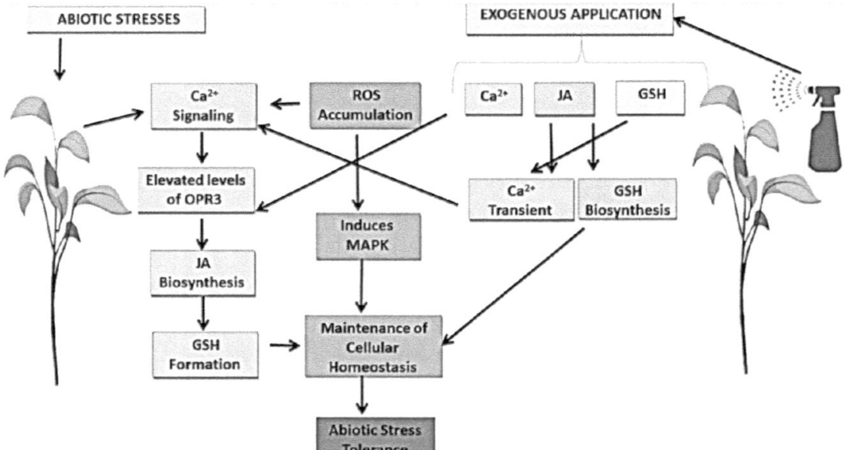

FIGURE 3.1 Signaling pathways involved in abiotic stress tolerance A. Abiotic stress causes ROS accumulation that leads to Ca^{2+} signaling and MAPK pathways to combat abiotic stress in plants; B. Exogenous application of JA, Ca^{2+} and GSH under stress condition mediates activation of Ca^{2+} signaling and MAPK pathways via possible cross-talk.

Source: Reprinted from Aslam et al. (2021). https://creativecommons.org/licenses/by/4.0/

Jasmonates are biosynthesized from lipid substrate, linolenic acid forming 12-OPDA using enzymes like allene oxide cyclase, 13-lipoxygenase, allene oxide synthase, etc. The first reactions happen in chloroplasts and the final production of JA occurs in peroxisomes by utilizing OPR3, ACX1. Finally, these are transported to cytosol; JA biosynthesis process is a positive-feedback homeostatic process. In the cytosol, JA-Ile is formed using the biocatalyst JAR1. Another mechanism suggests that in primitive plant classes, such as in liverworts and mosses, OPR3 is absent and continuous beta-oxidation in peroxisome results into formation of 5-didehydro-JA in such plants; further reduction to JA takes place using OPR2, an alternative to OPR3.

The JA signaling passageway is divided into two sub-divisions: MYC branch is based on transcription factors MYC2, MYC3 and MYC4. The MYC proteins interconnection with JAZ repressors further inhibits MED25 subunit interaction with them. ERF branch marker gene expressions, act against MYC mechanisms and synergistic to ethylene signaling pathway. In the necrotrophic defense process of plants, both jasmonates and ethylene are necessary.

Salicylic acid, a phenolic substance, on injecting into tobacco leaves, convey resistance to TMV; therefore, it has proven to perform a specific function of pathogenesis-resistant gene expression. SA-mediated defense is exerted by NPR1 action at post-transcriptional level. TGA family transcription factors also interact with NPR1 at different levels in presence of SA. Studies have shown that at low concentrations of JA and SA, both PR1 and PDF1.2 genes express in mutual way while in case of higher concentrations, they are expressed in antagonistic way. Cross-talks of both the pathways can be demonstrated by focusing on antagonistic expressions of SA and JA-responsive genes. When a biotroph (*Pseudomonas* syringae) attacks and in turn SA signaling pathway gets activated, susceptibility towards necrotrophs also increases due to repression of ET/JA signaling passage. Though this suppression is not dependent on biosynthesis of jasmonates, but it has been shown that SA introduction to the plant tissues can actually suppress JA-marker gene named PDF1.2 expression, similar to the AOS mutation. COR is a mimicking molecule for JA-Ile that binds to COI1-JAZ coreceptor to intensify the JA response and NAC transcription factor functional capabilities, restrains ICS1, activates BSMT1 and thus SA aggregation diminishes.

The molecular concept of SA/JA cross-talk is dependent on a few genes and transcription factors. WRKY TFs are somehow incriminated in SA induced responses and WRKY70 in Arabidopsis is the common point to explain the cross-talk between SA and JA signaling pathways. Overexpression of the mentioned factor in turn increases expression of SA related PR genes and represses JA linked gene PDF1.2; WRKY17 is a negative controller WRKY70. The importance of WRKY62 in SA/JA cross-talk has recently been implemented which shows that overexpression of the TF-WRKY62 inhibits the VSP2 and LOX2 genes, which can concomitantly be stimulated by MeJ (Thaler et al., 2012).

Glutaredoxin GRX480 is another regulator of SA/JA signaling interaction. Thiol disulfide reductions and redox activities of proteins have been regulated by these factors. Only a few of JA-mediated genes are affected by GRX480, which are sensitive to TGA and NPR1 modulations and SA-induced responses. NPR1 is a central regulator protein which on interaction with TGA factors results into PR genes. Cytosolic OsNPR1 when expressed in higher level increases the susceptibility to insect herbivory. During herbivore assault, this NPR1 factor suppresses SA/JA cross-talk, prioritizing JA responses to the attack and diminishing SA-responses. The substrate of MKS1 is phosphorylated by mitogen-activated protein kinase 4 and it suppresses SA-mediated signaling, whereas PAD4 represses JA-signalling (Koornneef & Pieterse, 2008).

3.2.4 JA CROSS-TALK WITH GA, ET, ABA, AND AUXIN

Gibberellin and ethylene are two phytohormones, which can be considered as secondary metabolites. Jasmonates cross-talk with both the chemicals in different conditions. DELLA proteins are a group of repressors for GA responses; the JAZ molecule, basic repressor for JA signaling, interacts with DELLAs and represses their activity, while in reciprocation, interacting DELLAs repress JAZ to maintain the susceptibility to *Pst* DC3000; JAZ degradation by JA signal transduction activation leads to PIF3 suppression in the GA mechanism, resulting in the inhibition of hypocotyl lengthening and DELLA degradation by GA signals decrease root elongation inhibitory effect. Not only antagonistically, but also these two phytohormones share signaling in different pathways in mutualistic manner; stamen development, trichome formation are also modulated by GA-JA cross-talks. Another secondary metabolite group called sesquiterpenes' anabolism is also influenced by JAZ-DELLA interaction with same transcription factors (Song et al., 2014).

Ethylene and jasmonates act independently for fungal resistance of the plant. Along with JA, reactive-oxygen species and ethylene too mediate secondary metabolites, when wounding or injury happens. For example, carrots produce a phenolic compound called caffeoylquinic acid as stress-response and there is an interrelation between the wound-induced synthesis of PCs and relative gene expression of ET-activity, JA and ROS biosynthesis-based genes under wounding stress. Data from studies have shown that after wounding happens, genes coding for ET and ROS biosynthesis get activated earlier. The rapid expression of ACC synthase gene, Glycolate oxidase gene signifies the production of ET and ROS. JA synthesizing gene gets activated comparatively later than the mentioned ones (Jacobo-Velázquez et al., 2015).

Other than the ET/JA and SA cross-talk in case of pathogenetic responses, under abiotic stress too ethylene and jasmonates cross-talk. In support of thermotolerance, both the chemicals act antagonistically; ethylene's function in case of sensitivity to heat-stress is dependent on strength of the triggering factor. Under ozone stress, JA protect the tissues instead of promoting cell-death. JA blocks ethylene pathway acting upon constitutive triple response kinase (CTR1). Both the chemicals' cross-talk in adventitious root formation occurs via EIN3/EIL1, induced by JAZ. In antagonistic cross-talk pathway, reactivation of CTR1 inhibits EIN3/EIL1, while in synergistic interaction, both the substances encourage ERF transcription through EIN3/EIL1 stimulation (Li et al., 2019).

ABA and JA signaling pathways work against herbivorous insect attack. Metabolic re-programmimg is done via ABA receptor PYLs, through JA-mediated passage, by forming PYL-JAZ complex. This is the activator of MYC2 transcriptional functions. Although, ABA degrades JAZ12 and thus their cross-talk balances between growth and defense systems.

Auxin is a progressive regulator of plant growth and development. Endogenous JA content stimulates the level of auxin by inducing higher expression of ASA1 and exogenous auxin can activate auxin-TIR-AUX/IAA-ARF signaling machinery and enhance JA production. ARF6/ARF8 in the auxin-response pathway, control flower (stamen and petal) growth through JA level modulation, while they act antagonistically in root growth regulation via JAZ-MYC2 mechanism (Pérez-Llorca et al., 2023; Yang et al., 2019) (Figure 3.2).

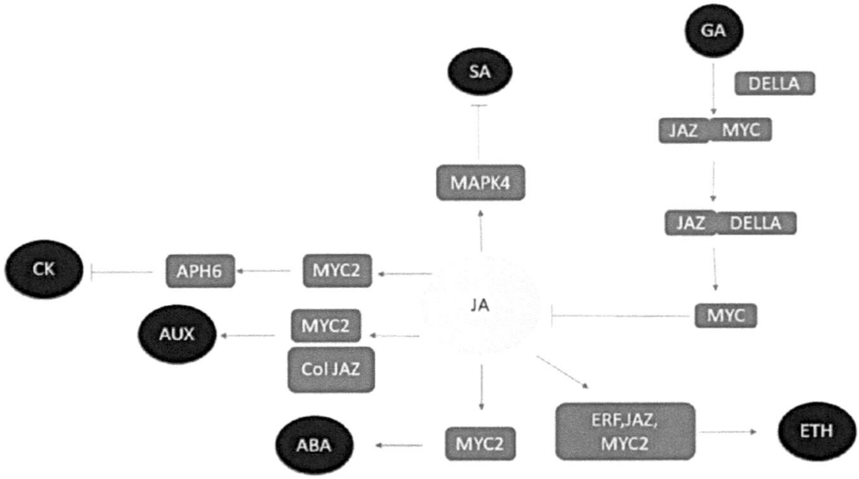

FIGURE 3.2 Cross-talk between jasmonates and phytohormones under different abiotic stress via transcription factors.

Source: Reprinted from Aslam et al. (2021). https://creativecommons.org/licenses/by/4.0/

3.3 ANTIOXIDANTS AND JA SIGNALING PATHWAY CROSS-TALK/ REGULATIONS

The interaction of JA and antioxidants can be described as in case of ascorbate too. Ascorbate reacts with reactive oxygen species and forms substrate for Asa peroxidase enzyme. When an Arabidopsis plant faces oxidative stress, Asa level decreases; upregulation of Asa pathway indicates constitutive response of JA-signaling. In JAs-depleted plants, low levels of NADP, UDP-glucose, higher level of sucrose, myo-inositol, quantitatively modified TCA cycle metabolites can be observed. These intermediates are the precursors of several secondary metabolites. The amount of sulfur-containing metabolites get reduced in AOS plants establishes the relation between JA and sulfur-based secondary metabolites' regulation.

Jasmonates and its derivatives regulate plant metabolic activities including flowering, seed germination, senescence and aging, root, and callus growth, etc. They either intensify or subdue the response under different conditions and have the capability of lifting up the antioxidant workings inside the plants. These signaling molecules act to defend biotic-abiotic stresses through cascade machineries, not alone but interacting with other molecules. Glutathione is the "master antioxidant" of plant body, present everywhere at high concentrations; this tripeptide thiol signifies the responses in chilling damage, salt-stress, functions of disease-resistance, cell division, redox-state determination of plant cells. These two biochemical substances, along with calcium ions, cross-talk and maintain tolerance against stresses via signaling cascades.

Different abiotic stresses which affect the productivity of the plants, are tolerated by the actions of JA, GSH and calcium ions. Industrial expansion and anthropogenic acts have resulted into overaccumulation and contamination of soil and water by several micronutrients. These conditions hamper photosynthesis, transpiration, and other biological processes of plants. Depending on the plants species which get exposed to metal stress, exogenous JA utilization varies and thus its impact on modulating primary metabolite production, diminishing biosorption of heavy metals. Exogenous calcium ion application stabilizes cell membrane physiology, blocks the heavy metal ion uptake, higher the soluble sugar and proline amount in cells-tissues. GSH is an important part of glyoxalase system too. Chelation and detoxification by GSH protect the plant from oxidative stress.

Jasmonates is of utmost importance in regulating antioxidant responses. JA potentially measures increased level of GSH and ROS scavenging with the help of enzyme-based antioxidants. Calcium signaling interacts with JA signaling using calcium channels, which get opened by ROS-aggregation and synthesis in response to abiotic stress. Calcium ion fluctuations governs JA biosynthesis and temporary calcium influx and level rise are mediated by JA and other derivates in cytoplasm and nucleoplasm; cytosolic calcium level is responsible for activation of CDPK, CAMs, in turn influencing stress-response mechanisms. CAM and CAM like proteins are involved in JA associated gene expression. CML42 in Arabidopsis have been reported to have indispensable impact on calcium mediated biological synthesis of jasmonates. ROS is sequestered by antioxidants and in response to JA, GSH related enzymes coding genes are also transcribed. JA enhances GCL-GS activity, further leading to glutathione disulfide or GSSG implementation. JA-induced signaling cascade are found to be upregulating gene expression linked to GSH metabolism. JA finds an application in nifedipine sensitive calcium ion channel openings, proving the cross-talk between JA-activated influx of calcium from extracellular spaces and inositol triphosphate-based calcium storages. Different level "Omics" studies have deciphered protein profile changes and fast-slow responding gene classifications. Hormones regulated gene responses, MYC2-MYC3 target genes, mutations in genes are also inked to JA response alterations; JA precursor 12-OPDA and GSTs cross-talk and elevate stress-tolerance as seen in tomato. Different omics approaches have revealed mechanisms of cross-talk among jasmonates and other stress-responding biochemicals at genomic, transcriptomic, metabolomic levels.

Increased ROS level and oxidative stresses as a result of nature-based and anthropogenic changes, can harm the primary metabolites and structural-functional features of plant tissues. A number of enzymes (peroxidase, catalase, superoxide dismutase, etc.) and other biomolecules (AsA, GSH, etc.) compose the group of antioxidants to protect the plant. In the reduction of hydrogen peroxide to water, ascorbic acid acts as a donor of electrons, AsA gets converted to MDHA, which can reversibly get reduced to AsA using MDHAR enzyme. AsA-GSH cycle helps in detoxification of ROS. Tocopherol, a type of vitamin E, works in terms of saving the ells from lipid peroxidation. Carotenoids, a type terpenoids, are also able to scavenge ROS, protect the plant from excess light stress by evaporating the unwanted energy as heat. Aromatic compounds called polyphenols are also antioxidants, containing hydroxyl groups. Anthocyanins are also photoprotective which play a major role in providing resistance to visible spectrum, while flavonoids work against UV-spectrum. Transgenic methods can be a solution to low-tolerance problems of plants to oxidative stressful conditions that can manipulate expression of genes related to biosynthesis of antioxidants and factors associated with the levels of antioxidant-property carrying secondary metabolites (Llauradó Maury et al., 2020; Aslam et al., 2021).

3.4 CONCLUDING REMARKS

Both phytohormones and secondary metabolites cooperatively work for the pathogen-herbivore resistance using regulatory pathways. Recent studies have shed light on novel molecular mechanisms and the flexible, controlling signal-networks of the pathways. Ecological perspective of plant defense-system has also come to highlights in recent years. Two responses have been focused- slowing down the plant growth-rate and enhancement of tolerance. Following the strategy of survival of the fittest, every organism, including the plants, wants to achieve maximum fitness even in the presence of challenges to the resources. JA-mediated regulation of secondary metabolic pathways, deciphered as cross-talks, are still too complex to elucidate in appropriate manner. Although, a few mechanisms, such as protein-protein interactions, gene-TF connections have been illustrated till now. The abiotic stresses are the major threats to the agricultural field and therefore the understanding of such cross-talk cascades, at biochemical, omics level has become an important strategy to deal with it. It is necessary to analyze the hormonal and secondary metabolite signaling transduction networks at each developmental stage of plant life. Better research on these mechanisms will help in quality enhancement of crops utilizing biotechnological engineering techniques in near future (Afrin et al., 2015; Jang et al., 2020).

KEYWORDS

- **antioxidant enzymes**
- **cross-talk**
- **defense mechanisms**
- **glucosinolate pathway**
- **jasmonates**
- **regulatory genes**
- **secondary metabolites**

REFERENCES

Afrin, S., Huang, J. J., & Luo, Z. Y. (2015). JA-mediated transcriptional regulation of secondary metabolism in medicinal plants. *Science Bulletin, 60*(12), 1062–1072. https://doi.org/10.1007/s11434-015-0813-0.

Aslam, S., Gul, N., Mir, M. A., Asgher, M., Al-Sulami, N., Abulfaraj, A. A., & Qari, S. (2021). Role of jasmonates, calcium, and glutathione in plants to combat abiotic stresses through precise signaling cascade. *Frontiers in Plant Science, 12*. https://doi.org/10.3389/fpls.2021.668029.

Guerriero, G., Berni, R., Muñoz-Sanchez, J. A., Apone, F., Abdel-Salam, E. M., Qahtan, A. A., Alatar, A. A., Cantini, C., Cai, G., Hausman, J. F., Siddiqui, K. S., Hernández-Sotomayor, S. M. T., & Faisal, M. (2018). Production of plant secondary metabolites: Examples, tips, and suggestions for biotechnologists. *Genes, 9*(6). https://doi.org/10.3390/genes9060309.

Iqbal, N., Masood, A., Khan, M. I. R., Asgher, M., Fatma, M., & Khan, N. A. (2013). Cross-talk between sulfur assimilation and ethylene signaling in plants. *Plant Signaling and Behavior, 8*(1), 104–112. https://doi.org/10.4161/psb.22478.

Jacobo-Velázquez, D. A., González-Aguëro, M., & Cisneros-Zevallos, L. (2015). Cross-talk between signaling pathways: The link between plant secondary metabolite production and wounding stress response. *Scientific Reports, 5*. https://doi.org/10.1038/srep08608.

Jang, G., Yoon, Y., & Choi, Y. Do. (2020). Crosstalk with jasmonic acid integrates multiple responses in plant development. *International Journal of Molecular Sciences, 21*(1). https://doi.org/10.3390/ijms21010305.

Jogawat, A., Yadav, B., Chhaya, Lakra, N., Singh, A. K., & Narayan, O. P. (2021). Crosstalk between phytohormones and secondary metabolites in the drought stress tolerance of crop plants: A review. *Physiologia Plantarum, 172*(2), 1106–1132. https://doi.org/10.1111/ppl.13328.

Koo, A. J. K., & Howe, G. A. (2009). The wound hormone jasmonate. *Phytochemistry, 70*(13, 14), 1571–1580. https://doi.org/10.1016/j.phytochem.2009.07.018.

Koornneef, A., & Pieterse, C. M. J. (2008). Cross talk in defense signaling. *Plant Physiology, 146*(3), 839–844. https://doi.org/10.1104/pp.107.112029.

Larrieu, A., & Vernoux, T. (2016). Q&A: How does jasmonate signaling enable plants to adapt and survive? *BMC Biology, 14*(1). https://doi.org/10.1186/s12915-016-0308-8.

Li, N., Han, X., Feng, D., Yuan, D., & Huang, L. J. (2019). Signaling crosstalk between salicylic acid and ethylene/jasmonate in plant defense: Do we understand what they are whispering? *International Journal of Molecular Sciences, 20*(3). https://doi.org/10.3390/ijms20030671.

Llauradó Maury, G., Méndez Rodríguez, D., Hendrix, S., Escalona Arranz, J. C., Fung Boix, Y., Pacheco, A. O., García Díaz, J., Morris Quevedo, H. J., Ferrer Dubois, A., Isaac Aleman, E., Beenaerts, N., Méndez Santos, I. E., Orberá Ratón, T., Cos, P., & Cuypers, A. (2020). Antioxidants in plants: A valorization potential emphasizing the need for the conservation of plant biodiversity in Cuba. *Antioxidants, 9*(11), 1–39. https://doi.org/10.3390/antiox9111048.

Pérez-Llorca, M., Pollmann, S., & Müller, M. (2023). Ethylene and jasmonates signaling network mediating secondary metabolites under abiotic stress. *International Journal of Molecular Sciences, 24*(6), 5990. https://doi.org/10.3390/ijms24065990.

Ramakrishna, A., & Ravishankar, G. A. (2011). Influence of abiotic stress signals on secondary metabolites in plants. *Plant Signaling and Behavior, 6*(11), 1720–1731. https://doi.org/10.4161/psb.6.11.17613.

Savchenko, T. V., Rolletschek, H., & Dehesh, K. (2019). Jasmonates-mediated rewiring of central metabolism regulates adaptive responses. *Plant and Cell Physiology, 60*(12), 2613–2620. https://doi.org/10.1093/pcp/pcz181.

Sohn, S. I., Pandian, S., Rakkammal, K., Largia, M. J. V., Thamilarasan, S. K., Balaji, S., Zoclanclounon, Y. A. B., Shilpha, J., & Ramesh, M. (2022). Jasmonates in plant growth and development and elicitation of secondary metabolites: An updated overview. *Frontiers in Plant Science, 13*. https://doi.org/10.3389/fpls.2022.942789.

Song, S., Qi, T., Wasternack, C., & Xie, D. (2014). Jasmonate signaling and crosstalk with gibberellin and ethylene. *Current Opinion in Plant Biology, 21*, 112–119. https://doi.org/10.1016/j.pbi.2014.07.005.

Thaler, J. S., Humphrey, P. T., & Whiteman, N. K. (2012). Evolution of jasmonate and salicylate signal crosstalk. *Trends in Plant Science, 17*(5), 260–270. https://doi.org/10.1016/j.tplants.2012.02.010.

Yang, J., Duan, G., Li, C., Liu, L., Han, G., Zhang, Y., & Wang, C. (2019). The crosstalks between jasmonic acid and other plant hormone signaling highlight the involvement of jasmonic acid as a core component in plant response to biotic and abiotic stresses. *Frontiers in Plant Science, 10*. https://doi.org/10.3389/fpls.2019.01349.

CHAPTER 4

Multifunctional Role of Jasmonates in Stress Tolerance

KAISAR AHMAD BHAT,[1,2] NAHILA ANJUM,[1] ARJUMAND JOHN,[3] ALI ASGHAR SHAH,[1] MOHD. HANIEF,[1] and SAJAD MAJEED ZARGAR[2]

[1]*School of Biosciences and Biotechnology, Baba Ghulam Shah Badshah University, Rajouri, Jammu and Kashmir, India*

[2]*Proteomics Laboratory, Division of Plant Biotechnology, Sher-e-Kashmir University of Agricultural Sciences and Technology of Kashmir (SKUAST-K), Shalimar, Jammu and Kashmir, India*

[3]*Department of Zoology, Lovely Professional University, Phagwara, Jalandhar, Punjab, India*

ABSTRACT

Plants are constantly exposed to a wide array of environmental stresses, including abiotic as well as biotic factors, which significantly affect their growth, development, and productivity. To cope with these challenges, plants have evolved complex defense mechanisms, where hormone jasmonate plays a key role in stress tolerance. It is mostly known that jasmonic acid (JA) is the most abundant and well-studied form of plant hormone, a group of lipid-derived signaling molecules. The role of jasmonates in abiotic stress responses such as drought, salinity, extreme temperatures, and oxidative stress cannot be overstated. In addition to regulating stomatal closure, enhancing antioxidant defenses, modulating osmotic adjustments, and promoting the accumulation of compatible solutes, they contribute to better tolerance to stress. A wide range of physiological and

Biotechnological Applications of Jasmonates: A Group of Less Well-Known Plant Growth Regulators.
Tariq Ahmad Dar, Bilal Ahmad, and Khalid Rehman Hakeem (Eds.)

biochemical changes are orchestrated by jasmonates through the regulation of several stress-responsive genes and transcriptional factors. Apart from abiotic stress, jasmonates play an important role in plant defense against biotic stress as well. By inducing defense-related genes, synthesizing antimicrobial compounds, and reinforcing physical barriers, they serve as key components in activating plant immunity. Accordingly, this book chapter highlights the multifunctional role that jasmonates in stress tolerance, including their involvement in abiotic stress responses, biotic stress defense, and integration with plant growth and development.

4.1 INTRODUCTION

Jasmonates are a group of lipid-derived hormones that regulate a wide range of physiological processes in plants, including growth, development, and stress responses (Wasternack and Kombrink, 2010). The most well-known and abundant form of jasmonates is jasmonic acid (JA). They are synthesized from α-linolenic acid, a polyunsaturated fatty acid present in plant cell membranes (Huan et al., 2022). The most well-known and extensively studied form of jasmonates is jasmonic acid (JA), but other derivatives such as methyl jasmonate (MeJA) and jasmonoyl-isoleucine (JA-Ile) also exist (Yan et al., 2013). Initially identified for its involvement in plant defense against herbivorous insects, JA has emerged as a multifunctional regulator in stress tolerance. It is now recognized that jasmonates, particularly JA, have a significant impact on both abiotic and biotic stress responses, making them key players in plant survival and adaptation (Wasternack and Hause, 2013). In the context of abiotic stress tolerance, jasmonates have been shown to mediate numerous adaptive responses. They participate in the regulation of stomatal closure, which reduces water loss and maintains leaf hydration during periods of drought and salinity stress (Yan et al., 2013). Jasmonates also contribute to the enhancement of antioxidant defense systems, mitigating the harmful effects of reactive oxygen species (ROS) generated under stress conditions. Furthermore, they modulate osmotic adjustments, facilitating the accumulation of compatible solutes that help maintain cellular turgor and protect against dehydration (Arif et al., 2020). Through these mechanisms, jasmonates play a critical role in improving plant tolerance to abiotic stresses. In addition to their involvement in abiotic stress responses, jasmonates are integral to plant defense against biotic stresses (Santino et al., 2013). They function as key components in the activation of plant immune responses to pathogens and herbivores. Jasmonate signaling triggers the production of antimicrobial compounds, such as phytoalexins, which inhibit the growth of pathogens, limiting their spread within plant tissues (Rojo et al., 2003). Jasmonates also induce the synthesis of defense-related proteins, protease inhibitors, and secondary metabolites that deter or directly harm herbivores, acting as an effective defense strategy against insect attacks (Divekar et al., 2022). The interplay between jasmonate signaling and other hormone pathways, such as salicylic acid and ethylene, further fine-tunes the plant's defense responses against biotic stressors (Nguyen et al., 2016). Understanding the multifunctional role of jasmonates in stress tolerance is of great significance in the field of plant biology and agricultural sciences. By deciphering the intricate regulatory mechanisms of jasmonate signaling, researchers can develop targeted strategies to enhance stress tolerance in crops (Raza et al., 2019). Genetic engineering and targeted agronomic interventions aimed at manipulating jasmonate pathways hold immense potential for the development of stress-tolerant crop varieties, thus contributing to global food security and sustainable agriculture (Mishra and Sharma, 2022).

4.2 JASMONATE BIOSYNTHESIS AND METABOLISM

In monocotyledonous and dicotyledonous plants, JA biosynthesis is controlled by several enzymes (Han, 2017), and it has been extensively studied since the 1950s (Ali and Beak, 2020). Chloroplast membranes synthesize JA from polyunsaturated fatty acids (Schaller and Stintzi, 2009). Despite the fact that certain plants use 16-carbon fatty acids for the biosynthesis of JA, most plants use 18-carbon fatty acids. The enzyme 13-lipoxygenase (LOX) converts the 18-carbon precursor into 13(S)-hydroperoxy-octadecatrienoic acid (13-HODE). Afterward, allene oxide synthase (AOS) and allene oxide cyclase (AOC) catalyze the 13-HODE reaction (Han, 2017). In the process of dehydrating 13-HODE, AOS produced an unstable allylic epoxide (allene oxide), 12,13(S)-epoxy-octadecatrienoic acid (12,13-EOT) (Weber et al., 1997). Peroxisomes import OPDA via ABC transporters PXA1, PED3, or COMATOSE (CTS). In the peroxisome, OPDA reductase 3 (OPDA 3) reduces OPDA, and then the reduced OPDA is activated by OPC-8:0 CoA Ligase1 (OPCL1). The last step is β-oxidation, which results in the formation of the 12-carbon molecule JA by removing carbons from the carboxyl side chain (Han, 2017). Three core enzymes are involved in β-oxidation, acyl-CoA oxidase (ACX), a multifunctional protein having 2-trans-enoyl-CoA hydratase,3-ketoacyl-CoA thiolase (KAT), and L-3-hydroxy acyl-CoA dehydrogenase (ACH) (Li et al., 2005). JA is then transported to the cytosol, where it is converted into over 30 active and inactive derivatives depending on the chemical alteration of the carboxylic acid group, pentanone ring, or the pentenyl side chain (Figure 4.1) (Ali and Baek, 2020). Evidence shows that JA-Ile is a key component in the JA signal transduction pathway (Wasternack and Song, 2017). Other modifications of JA include methyl esterification, which results in the production of methyl jasmonate (MeJA), by the activity of JA-methyl transferase (JMT) (Han, 2017). In order to convert MeJA and JA-Ile to JA, two enzymes are required MeJA esterase (MJE), and JA-Ile hydrolase 1 (JIH1) (Koo et al., 2013; Woldemariam et al., 2012). JA is decarboxylated to form cis-jasmone (CJ) (Koch et al.,1997).

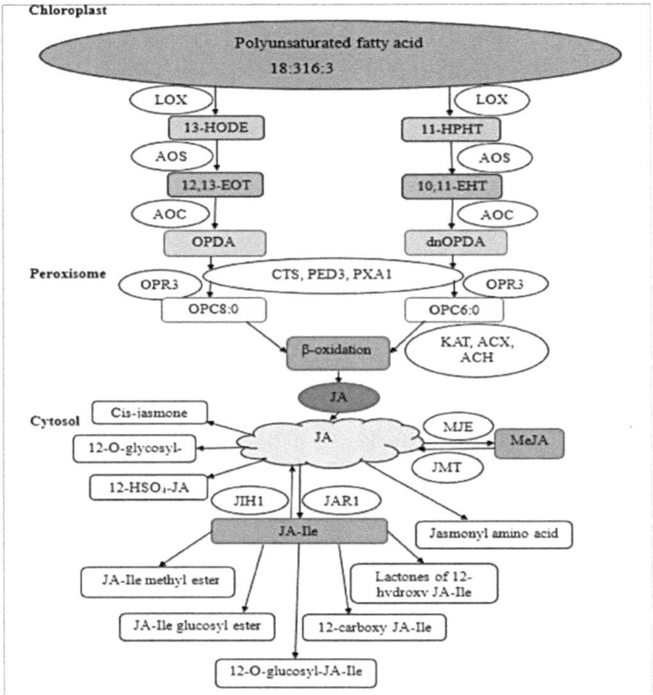

FIGURE 4.1 Jasmonate biosynthesis and metabolism: The sequential action of plastid-localizedLOX, AOS, and AOC results in the conversion of polyunsaturated fatty acids (18:3 and 16:3) to OPDA and dnOPDA in the chloroplast. Upon transport of OPDA and dnOPDA into peroxisomes, OPR3 and three cycles of β -oxidation result in the formation of JA, which is then transported to the cytoplasm. JA undergoes several modifications including JA-Ile and MeJA with the help of JAR1 and JMT, respectively.

4.3 ROLE OF JASMONATES IN ABIOTIC STRESS TOLERANCE

There are several biotic and abiotic stresses that plants experience as they grow in their environments. Abiotic stresses such as temperature, light, salt, water, carbon dioxide, oxygen, ozone, and soil nutrient availability affect various physiological processes in plants. The static nature of plants prevents them from switching to a better habitat quickly in order to avoid abiotic stresses. In order to cope with unfavorable environments, they have developed systems to adjust their physiological and developmental processes (Ali and Baek, 2020). As plant development regulators, phytohormones serve a variety of functions, including sensing drought, osmotic stress, heavy metal toxicity, chilling injury, and other environmental conditions (Lymperopoulos et al., 2018). Natural phytohormones include nine groups (auxins, cytokinins, gibberellins, abscisic acid, ethylene, JA, brassinosteroids, salicylic acid, strigolactones, etc.), all of which evoke a variety of responses in the environment (Ali and Baek, 2020). JA ZIM domain (JAZ) proteins inhibit transcription factors under favourable environmental conditions or at low levels of JA-Ile, preventing them from activating JA-responsive gene promoters. As soon as the JA-Ile reaches the nucleus, it is transported by the JA transfer protein 1 (JAT1). The JAT1 enzyme modulates the cytoplasmic, nucleolar, and apoplastic dynamics of JA during abiotic stress. Other cells are stimulated by JA signaling pathways when JA or JA-Ile is present in the apoplast. It is possible for JA signals to travel long distances through vascular bundles or air transmissions. The high volatility and ability to penetrate cell membranes of MeJA make it capable of quickly dispersing to distant leaves and neighboring plants as compared to JA. Within the SCF complex, JA-Ile promotes the association of JAZ with the F-box protein coronatine insensitive 1 (COI1), resulting in the destruction of JAZ by proteasomes. As a result of the activation of the mediator (MED25), RNA polymerase II, and general transcription factor (GTF) by the derepressed transcription factor (TF), JA-responsive genes begin to be expressed (Ali and Beak, 2020; Farmer and Ryan, 1990). A high concentration of JA is found in reproductive tissues and flowers of higher plant species, whereas low concentrations are found in mature leaves and roots (Dar et al., 2015; Wasternack and Home, 2013). Plant defense systems are also activated by JAs in response to insect wounding and pathogen assault (Ali and Bake, 2020; Sharma and Laxmi, 2016) (Figure 4.2).

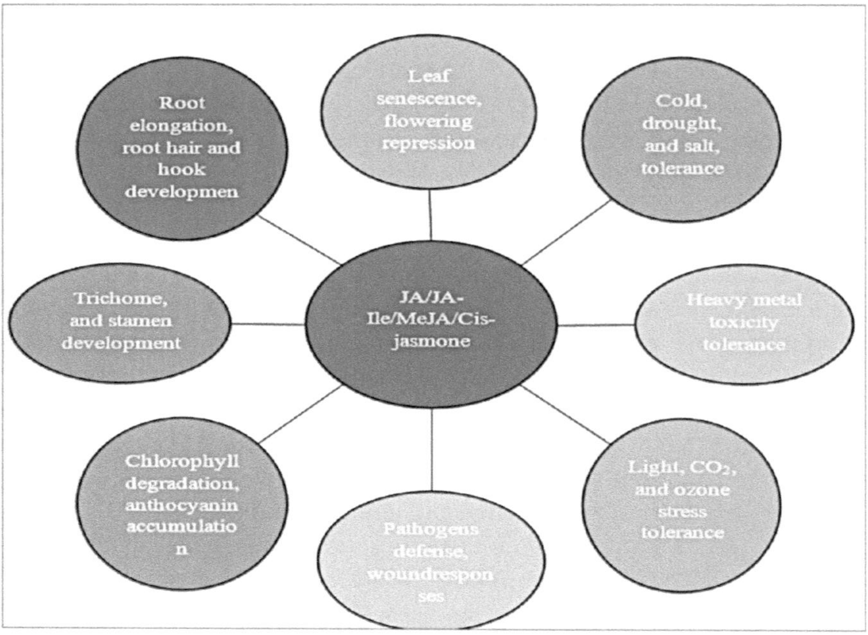

FIGURE 4.2 Numerous plant processes are controlled by JA and its isoleucine conjugate in response to abiotic stresses.

4.3.1 DROUGHT TOLERANCE

Among all environmental factors, drought is perhaps the most significant barrier to crop quality and production, threatening economic output and the availability of food for people (Roche et al., 2007). The main effects of drought on crop plants include reduced cell division and growth rates, smaller leaves, longer stems and roots, and disturbed stomatal oscillations. These effects also have an impact on plant water and nutrient relationships, which lower crop productivity, and water use efficiency (WUE) (Li et al., 2009; Farooq et al., 2009). Abscisic acid (ABA) biosynthesis is accelerated by drought, which decreases stomatal conductance to reduce transpirational losses (Yamaguchi-Shinozaki and Shinozaki, 2006). Plants may adapt to water stress by changing their cellular metabolism to activate several defense systems (Abdelgawad et al., 2014). Proline, glycine betaine, sugars, and certain inorganic ions are examples of small compatible solutes (osmolytes) that are synthesized and accumulated in this process (Chaves et al., 2003; Ramanjulu and Bartels, 2002). Phytohormones are a part of signaling pathways, or their presence may activate signaling reactions or molecules that allow plants to respond to stresses. Jasmonates, a class of phytohormones produced by the metabolism of membrane fatty acids, has drawn a lot of interest (Avanci et al., 2010; Ueda et al., 2006; Norastehnia et al., 2007; Lee et al., 1996). Exogenously applied JAs stimulate a variety of physiological responses to stress, increasing plant resistance. MeJA promotes maize development under drought stress by prompting plants to produce more organic Osmo protectants (proline, free amino acids, and total soluble sugars), specific growth hormones, and oxidative enzymes (superoxide dismutase (SOD), peroxidase (POD), and catalase (CAT)) (Abdelgawad et al., 2014). Jasmonates were also used to increase the drought resistance of sugar beet (Beet vulgaris), soybean (Glycine max), and barley (Hordeum vulgare). When plants are exposed to osmotic stress caused by polyethylene glycol and water retention, MeJA application can increase their tolerance to dehydration. MeJA regulates a wide range of plant development processes, such as seed germination, root development, flowering, fruit ripening, senescence, and resistance to abiotic stresses (Ndiaye et al., 2022).

4.3.2 COLD STRESS TOLERANCE

Cold stress, which includes chilling (below 20°C) and/or freezing (below 0°C) conditions, has a negative impact on plant growth and development which severely reduces plant distribution and agricultural production. Due to its direct inhibition of metabolic processes and indirect effects on osmotic (cellular dryness caused by freezing and inhibition of water intake caused by chilling), oxidative, and other stresses, cold stress hinders plants from expressing their full genetic potential (Chinnusamy et al., 2007). Plants have developed a variety of physiological, biochemical, and molecular tolerance mechanisms as a response to cold stress, allowing them to endure harsh climatic circumstances. (Shinozaki et al., 2003). Thus, it was found that after cold stress, wheat had higher levels of JA and SA. Additionally, when a population of Pinus *pinaster* plants was exposed to drought and cold temperatures, their JA content increased several times (Kolaksazov et al., 2013). Moreover, cold stress significantly influenced the production of LOX, in both kiwi (*Actinidia delicosa*) and *Caragana jubata* (Zhang et al., 2006; Bhardwaj et al., 2011). Cold treatment leads to a rise in JA accumulation in *Arabidopsis*, which causes the degradation of JA signaling repressors, JAZ1, and JAZ4 proteins, which interact and suppress the ICE1-CBF module. As a result, JA increases the ability of *Arabidopsis* ability to withstand cold temperatures by activating the ICE1-CBF cascade (Hu et al., 2013). The ability of rice to withstand low temperatures is positively regulated by JA, as HAN1 decreases chilling tolerance by encoding an oxidase that converts active JA-Ile to inactive 12-OH-JA-Ile (Ding et al., 2022). Recent studies indicate that JA also provides cold resistance in a variety of horticulture plant species. In apples (*Malus hupehensis*) (An et al., 2020). In the banana (Musa acuminata), MYC2, a positive-JA signaling regulator, interacts with ICE1 to promote cold tolerance. It has also been demonstrated that JA promotes the production of osmolytes, hence enhancing cold tolerance (Ding et al., 2022).

4.3.3 HEAVY METAL TOXICITY TOLERANCE

Heavy metals (HMs) are naturally occurring elements with a high atomic number and at least 5 times greater density as compared to water (Tchounwou et al., 2012). HMs toxicity has several negative consequences on plant health, including growth inhibition, reduced biomass output, chlorosis of leaves, nutritional and water imbalances, and finally leaf senescence and plant mortality. HMs can disrupt gene expression, the synthesis of various proteins and secondary metabolites, hormonal signaling changes, and antioxidant defense functions. JAs globally normalize plant growth and development under HMs toxicity (Raza et al., 2021). Exogenous MeJA could protect European Black nightshade (*Solanum nigrum*) (Yan et al., 2015), *Arabidopsis thialana* (Maksymiec and Krupa, 2007), soybean (*Glycine max*) (Poonam et al., 2013), chili pepper (*Capsicum frutescens*) (Yan et al., 2013), and mangrove (*Kandelia obovate*) plants against cadmium (Cd)-induced damage by enhancing the activities of SOD, CAT, and ascorbate peroxidase (APX). Under Boron (B) stress, H_2O_2 and O_2^- were seen to rise, while peroxidase (POD) and SOD activities were raised by JA treatment, indicating that JA promoted reactive oxygen species (ROS) removal in *Puccinellia tenuiflora* caused by excess B and consequently decreased oxidative damage (Zhao et al., 2019). MeJA also reduced B toxicity in sweet wormwood (*Artemisia annua*) via lowering lipid peroxidation and promoting the synthesis of antioxidative enzymes (Ali and Beak, 2020).

4.3.4 SALT STRESS TOLERANCE

Stress from salt is a major environmental factor that affects plant growth. When salt stress occurs, intracellular osmotic pressure rises and sodium accumulates in harmful amounts (Zhao et al., 2011). As a result of salt stress, seeds do not germinate, they do not develop, and they do not flower or fruit (Quan et al., 2007; Park et al., 2013). Plants cannot absorb water and nutrients from salty soil due to the high sodium concentration. The lack of water and nutritional imbalance causes osmotic and ionic stress. As a result of these main stresses, oxidative stress can lead to a cascade of secondary stresses. One of the ways for combating salt stress in plants is the use of phytohormones, the most important of which are SA, ABA, JAs, and BRs. There is growing evidence that JAs help plants to cope with salt stress. Exogenous MeJA treatments can considerably decrease salt stress symptoms in soybean seedlings (Yoon et al., 2009). JAs generate salt stress-responsive protein in rice roots. (Ahmad et al., 2019). Dry mass production in rice cultivars is increased by exogenous JA spraying, and salt stress can be partially reduced by JA application (Dar et al., 2015). The MeJA pretreatment of *Pisum sativum* also relieved salt stress and increased CO_2 fixation rates. It has been reported that barley seedlings treated with JAs are more tolerant to salt stress. In response to salt stress, the amount of JA in Iris hexagonal leaves and rice roots increases (Ahmad et al., 2019). JA levels were dramatically elevated under salt stress to compensate for the effects of salt (Zhang et al., 2017) in a transcript profile analysis of stressed sweet potatoes. By enhancing APX and SOD activities, exogenous MeJA enhanced the salt resistance of the black locust tree (Robinia pseudoacacia) (Jiang et al., 2016).

4.3.5 HEAT STRESS

As a result of the greenhouse effect, future instances of high temperatures are predicted to occur more frequently. Increasing temperatures have been a significant constraint on agricultural output (Qu et al., 2013). Under heat stress, the functions of the tapetal cells are lost throughout the reproductive development stage, and the anther is dysplastic (Jagadish et al., 2010). To resist heat stress, plants have developed a variety of physiological and molecular mechanisms (Qu et al., 2013). Inwild-type (WT) *Arabidopsis*, the importance of JA signaling in thermotolerance has recently been demonstrated. Electrolyte leakage experiments indicated that exogenous application of low concentrations of MeJA preserved cell viability in heat-stressed plants. Furthermore, heating WT Arabidopsis resulted in the accumulation of numerous JAs including as JA-Ile, JA, MeJA, and OPDA. Furthermore, several of the proteins generated by JA are comparable to those induced by heat shock (Ding et al., 2001). However, no evidence was revealed that MeJA-induced thermotolerance enhanced heat shock protein (HSP) gene expression (Clerk et al., 2009; Sharma and Laxmi, 2015). High temperatures enhance the expression of small HSPs in tomatoes, which is consistent with the findings of Kadyrzhanova et al. 1998, who discovered that 42°C heat treatment elevated HSP 17.6 expression (Ding et al., 2001). Furthermore, exogenous MeJA treatment increased the activities of CAT, APX, POD, and SOD, improving the scavenging ability of ROS and decreasing the oxidative damage produced by heat stress. Heat stress and exogenous MeJA increased transcript levels of associated genes (*LpAOC*, *LpLOX2*, *LpJMT*, and *LpOPR3*) in the JA biosynthesis pathway, which might increase JA and MeJA content accumulation.

4.4 ROLE OF JASMONATES IN BIOTIC STRESS TOLERANCE

Jasmonic acid (JA) plays a crucial role in plant defense against biotic stress, which refers to stress caused by living organisms such as pathogens (e.g., bacteria, fungi, viruses) and herbivores (e.g., insects, nematodes) (Gimenez et al., 2019). When plants are subjected to biotic stress, they activate a complex signaling network that involves various defense mechanisms, and the jasmonic acid signaling pathway is an important component of this network (Verma et al., 2016).

4.4.1 INSECT RESISTANCE

Molecular signals control plant defense responses to insects, with JA being the most significant signaling molecule (Zhang et al., 2017). Numerous herbivorous insects commonly cause harm to plants, which causes endogenous JA levels to rise quickly (Al-Zahrani et al., 2020).

Interestingly, despite JA's non-toxicity to insects, plants sprayed with JA develop secondary metabolites and systemic resistance to prevent insect harm (Zhang et al., 2020; Qi et al., 2020). Attacks by herbivores usually cause plants to accumulate protective proteins that alter the digestive physiology of pests and shield plants from herbivory (Chen et al., 2017). JAs could cause these protective proteins to be produced. For instance, exogenous JA or MeJA can trigger the crucial defense enzyme polyphenol oxidase (PPO), which can successfully shield plants against herbivores (Tan et al., 2012). MeJA increased the synthesis of harmful proteins in maize and triggered plant defenses against Ostrinia furnacalis (Zhang et al., 2015). In addition, a variety of secondary metabolites produced by jasmonates and resulting from intricate metabolic pathways that branch out actively contribute to plants' insect resistance. Secondary metabolites, namely tannins, total phenols, total flavonoids, and lignin, have been found to increase resistance to herbivorous insect assaults during the past 10 years (Yan et al., 2021; Li et al., 2021; Xu et al., 2021). Although it is commonly acknowledged that JAs are essential for triggering plant defenses against herbivorous insects, it is still unknown how plants initially activate JA production in response to herbivore interactions. JAV1-JAZ8-WRKY51 (JJW), a new complex formed by JAV1-associated VQ domain protein 1 (JAV1), JAZ8, and WRKY51 in healthy plants, suppresses the expression of JA biosynthetic genes. As a result, JA can keep its basal level low to facilitate typical plant development. However, the calmodulin-dependent phosphorylation of JAV1 is induced in response to calcium influx when plants are harmed by insect assaults. For plant defense, the JJW complex is broken down and JA biosynthesis is initiated, which causes the amount of JA to rise quickly (Xu et al., 2021). *OsAOS1* and *OsAOS2* are both engaged in the production of JA that is stimulated by herbivores and are crucial in influencing the susceptibility of rice to herbivore eating (Yan et al., 2018). Increased wound-induced JA production from tomato TomLox-D overexpression strengthens the plant's defense against insect herbivory (Zeng et al., 2021). In tea plants, the JA biosynthesis gene *CsOPR3* is crucial for defense against herbivorous insects (Yan et al., 2013). Major regulators of Arabidopsis resistance to a generalist herbivore are MYC2, MYC3, and MYC4 (Xin et al., 2017; Ye et al., 2012). JAZ proteins, however, have reportedly lately been identified as candidates for immunological signaling interference by insect effectors (Figure 4.3).

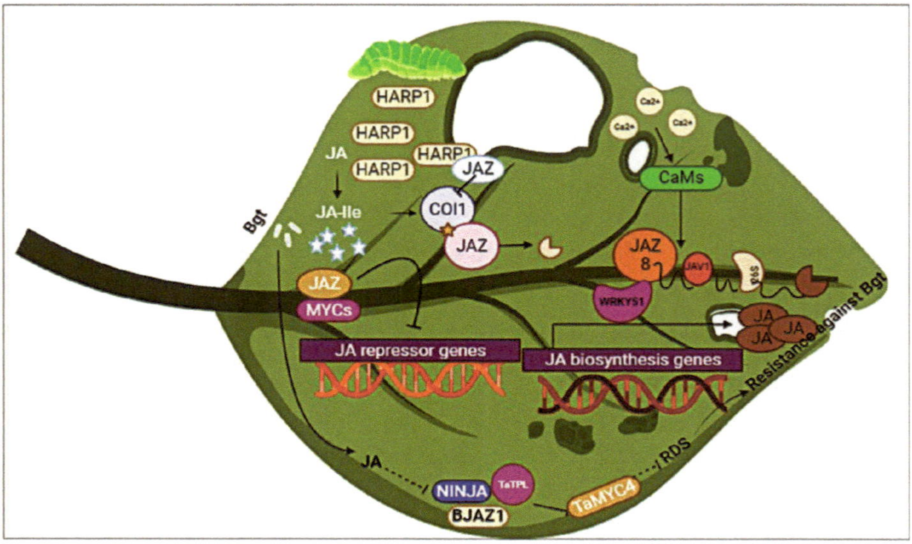

FIGURE 4.3 JA signaling in plant biotic stress response: The JA signaling pathway is activated when wheat is attacked by insects or pathogens. JA buildup can cause TaJAZ1 to degrade when exposed to Bgt, releasing TaMYC4, which prevents TaPRs expression and reactive oxygen species production, thus making bread wheat resistant to Bgt. The chewing insect effector *Helicoverpa armigera* HARP1 accumulates in leaves, inhibits JA-responsive gene expression, and reduces the plant's resistance to insect damage. A sudden burst of JA for plant defense occurs when JAV1 is Ca^{2+}/calmodulin-dependently phosphorylated during insect herbivory, which leads to the de-repression of JA biosynthesis genes.

4.4.2 PLANT DISEASE RESISTANCE

Endogenous signaling molecules known as JAs have a definite defensive impact on necrotrophic diseases in addition to being implicated in plants' insect resistance mechanism (Schweizer et al., 2013). Exogenous administration of JA or MeJA can stimulate the expression of plant defense genes, according to earlier research. Similarly, JAs therapy greatly increased the expression of multiple genes in the JA signaling system, enhancing resistance to necrotrophic infections (Chen et al., 2018). To protect wheat against Fusarium gramine arum, MeJA treatment resulted in a considerable elevation of PR4, PR5, and PEROXIDASE (PEROX) (Wang et al., 2012). Secondary metabolites and defensive enzymes undergo substantial modifications, which are crucial for plant defense against pathogen invasion (Ameye et al., 2015). In addition, exogenous JA treatment significantly raised the activities of peroxidase and PPO in citrus fruit, which can effectively suppress the occurrence of green and blue molds (Moosa et al., 2019). Application of MeJA may greatly reduce the level of H2O2 contents and lipid peroxidation in wheat infected by Fusarium culmorum (Zhao et al., 2021; Motallebi et al., 2017). By amplifying callose deposition, OPDA, a crucial step in the JA biosynthesis pathway, contributes to plant defense against Botrytis cinerea (Liu et al., 2019). JA-biosynthetic acyl-CoA oxidase (ACX) activity is promoted by CATALASE2 (CAT2) to increase plant tolerance to the necrotrophic Botrytis cinerea (Scalschi et al., 2015). Additionally, GhCOI1 is a crucial signaling element that regulates the JA-regulated defense response against *Alternaria brassicicola* in Gladiolus plants (Yuan et al., 2017), and earlier mechanical stress exposure of Arabidopsis seedlings improved JA-mediated defense against necrotrophic infections (Brenya et al., 2020). As a result of the biotrophic fungus *Blumeria graminis f.* sp. *tritici* (Bgt), bread wheat (*Triticum aestivum L.)* is frequently infected with powdery mildew, but transgenic bread wheat lines expressing *TaJAZ1* fragment without the Jas motif showed an enhanced expression of the pathogenesis-associated genes *TaPR1/2*. In the meantime, it was found that the degradation of TaJAZ1 caused *TaNINJA*, *TaTPL*, and *TaJAZ1* to form a co-repressor complex that limited TaMYC4's transcriptional activity (Seng et al., 2020). As a result of overexpression of the JA-related *OsbHLH034* gene, rice became hypersensitive to JA and adapted to rice bacterial blight more efficiently (Jing et al., 2019). As a result of its interaction with GhbHLH171, GhJAZ2 reduced JA-mediated defense against this fungus *Verticillium dahliae* in cotton (Onohata et al., 2020). In contrast to JAZ8 overexpression in Arabidopsis, which suppressed plant defense responses to necrotrophic diseases, JAZ8 suppressed the transcriptional activity of WRKY75 (Chen et al., 2020). These findings show that genes associated with JA signaling might enhance resistance to necrotrophic infections. In conclusion, endogenous and exogenous JAs both contribute significantly to how plants react to biotic and abiotic stress (Chen et al., 2020). Table 4.1 enlists the protective functions of JAs in response to stressors from insects, diseases, cold, salt, heat, and drought.

TABLE 4.1 Mechanism of JA Regulation in Response to Abiotic and Abiotic Stresses

Type	Stress	Plant Species	Jas	Protective Role
Abiotic stress	Drought	*Oryza sativa*	Endogenous	OsJAZ9 over-expression decreased the breadth and stomatal density of leaves
		Triticum sativum	Exogenous (JA)	Increased certain osmoregulation chemical accumulation, controlled the activity of antioxidant enzymes, and modulated morphological changes by restoring the shoot/root ratio
	Cold	*Oryza sativa*	Exogenous (JA)	Favorably affected the expression of numerous JA-related genes, including the AOC, AOS1, AOS2, LOX2, COI1a, and bHLH148
		Capsicum annuum	Exogenous (MeJA)	Increased endogenous JA content and antioxidant metabolite contents (glutamic acid, sucrose, and galactose), as well as genes associated with JA production being activated
	Salt	*Oryza sativa*	Endogenous	OsJAZ9 and OsbHLH062 interacted to control ion homeostasis
	Heat	*Oryza sativa*	Exogenous (MeJA)	Increased the number of JAs, improved the ability of leaves to respond to changes in osmotic pressure, and promoted spikelet blooming
		Solanum lycopersicum	Exogenous (JA)	JA/COI1 signaling pathway regulation rescued tomato stigma exertion
		Solanum lycopersicum	Endogenous	TomloxD over-expression improved heat tolerance
	Heavy metal (Ni)	*Glycine max*	Exogenous (JA)	Improved osmolytes, antioxidant enzyme activity, and gene expression
Biotic stress	*Ostrinia furnacalis*	*Zea mays*	Exogenous (MeJA)	Increased synthesis of harmful proteins and induced plant defense mechanisms
	Chilo suppressalis and *Niaparvata lugens*	*Oryza sativa*	Endogenous	OsAOS1 and OsAOS2 both participate in JA biosynthesis brought on by herbivores
	Cnaphalocrocis medinalis and *Nilaparvata lugens*	*Oryza sativa*	Endogenous	A trypsin protease inhibitor (TrypPI), POD, and PPO activation are all dependent on COI1
	Fusarium culmorum,	*Triticum aestivum*	Exogenous (MeJA)	Significantly reduced lipid peroxidation and H_2O_2 concentration levels

4.5 CONCLUSION

In conclusion, jasmonates, particularly jasmonic acid (JA), play an important role in plant stress tolerance. They are involved in diverse processes, including abiotic stress responses, biotic stress defense, and integration with plant growth and development. Jasmonates contribute to abiotic stress tolerance by regulating stomatal closure, enhancing antioxidant defense systems, modulating osmotic adjustments, and promoting the accumulation of compatible solutes. These processes support a plant's ability to endure challenging environmental factors such as oxidative stress, drought, salt, and severe temperatures. JAs have been the subject of a significant and thorough study on the regulation of stress-induced plant growth and defense since they are the primary hormones regulating biotic and abiotic stress responses. Model plants like Arabidopsis and rice have been used over time to gradually describe the routes for JA production, signaling, and metabolism, serving as crucial references for other plants. Although biotic and abiotic stressors have had significant negative economic effects on the production of certain significant crops and economic trees, few studies of these plants have been done. Therefore, more study on JAs in significant commercial plants is required as fundamental science to enable practical use against plant stresses. Understanding the multifaceted functions of jasmonates in stress tolerance has significant implications for agriculture and crop improvement. Manipulating jasmonate signaling pathways through genetic engineering and targeted agronomic interventions holds promise for enhancing stress tolerance in crop plants, leading to improved productivity and sustainability. Furthermore, further research into the complex regulatory mechanisms of jasmonate signaling will provide valuable insights and pave the way for innovative strategies to enhance stress tolerance in plants.

KEYWORDS

- **antimicrobial compounds**
- **biosynthesis**
- **biotic stress**
- **jasmonate biosynthesis**
- **jasmonates**
- **metabolism**
- **salinity stress**

REFERENCES

Abdelgawad, Z. A., Khalafallah, A. A., & Abdallah, M. M. (2014). Impact of methyl jasmonate on antioxidant activity and some biochemical aspects of maize plant grown under water stress condition. *Agricultural Sciences, 5*(12), 1077.

Ahmad, B., Raina, A., Naikoo, M. I., & Khan, S. (2019). Role of methyl jasmonates in salt stress tolerance in crop plants. In *Plant Signaling Molecules* (pp. 371–384). Woodhead Publishing.

Al-Zahrani, W., Bafeel, S. O., & El-Zohri, M. (2020). Jasmonates mediate plant defense responses to *Spodoptera exigua* herbivory in tomato and maize foliage. *Plant Signaling & Behavior, 15*(5), 1746898.

Ameye, M., Audenaert, K., De Zutter, N., Steppe, K., Van Meulebroek, L., Vanhaecke, L., & Smagghe, G. (2015). Priming of wheat with the green leaf volatile Z-3-hexenyl acetate enhances defense against *Fusarium graminearum* but boosts deoxynivalenol production. *Plant Physiology, 167*(4), 1671–1684.

An, J., Wang, X., Zhang, X., You, C., & Hao, Y. (2020). Apple B-box protein BBX37 regulates jasmonic acid-mediated cold tolerance through the JAZ-BBX37-ICE1-CBF pathway and undergoes MIEL1-mediated ubiquitination and degradation. *New Phytologist, 229*(5), 2707–2729.

Anjum, S. A., Wang, L. C., Farooq, M., Hussain, M., Xue, L. L., & Zou, C. M. (2011). Brassinolide application improves the drought tolerance in maize through modulation of enzymatic antioxidants and leaf gas exchange. *Journal of Agronomy and Crop Science, 197*, 177–185.

Arif, Y., Singh, P., Siddiqui, H., Bajguz, A., & Hayat, S. (2020). Salinity induced physiological and biochemical changes in plants: An omic approach towards salt stress tolerance. *Plant Physiology and Biochemistry, 156*, 64–77.

Avanci, N. C., Luche, D. D., Goldman, G. H., & Goldman, M. H. S. (2010). Jasmonates are phytohormones with multiple functions, including plant defense and reproduction. *Genetics and Molecular Research, 9*(1), 484–505.

Bhardwaj, P. K., Kaur, J., Sobti, R. C., Ahuja, P. S., & Kumar, S. (2011). Lipoxygenase in *Caragana jubata* responds to low temperature, abscisic acid, methyl jasmonate, and salicylic acid. *Gene, 483*(1, 2), 49–53.

Brenya, E., Chen, Z. H., Tissue, D., Papanicolaou, A., & Cazzonelli, C. I. (2020). Prior exposure of *Arabidopsis* seedlings to mechanical stress heightens jasmonic acid-mediated defense against necrotrophic pathogens. *BMC Plant Biology, 20*(1), 1–16.

Chaves, M. M., Maroco, J. P., & Pereira, J. S. (2003). Understanding plant responses to drought—From genes to the whole plant. *Functional Plant Biology, 30*(3), 239–264.

Chen, C. Y., Liu, Y. Q., Song, W. M., Chen, D. Y., Chen, F. Y., Chen, X. Y., & Mao, Y. B. (2019). An effector from cotton bollworm oral secretion impairs host plant defense signaling. *Proceedings of the National Academy of Sciences, 116*(28), 14331–14338.

Chen, H., Gonzales-Vigil, E., Wilkerson, C. G., & Howe, G. A. (2007). Stability of plant defense proteins in the gut of insect herbivores. *Plant Physiology, 143*(4), 1954–1967.

Chen, L., Zhang, L., Xiang, S., Chen, Y., Zhang, H., & Yu, D. (2021). The transcription factor WRKY75 positively regulates jasmonate-mediated plant defense to necrotrophic fungal pathogens. *Journal of Experimental Botany, 72*(4), 1473–1489.

Chinnusamy, V., Zhu, J., & Zhu, J. K. (2007). Cold stress regulation of gene expression in plants. *Trends in Plant Science, 12*(10), 444–451.
Clarke, S. M., Cristescu, S. M., Miersch, O., Harren, F. J., Wasternack, C., & Mur, L. A. (2009). Jasmonates act with salicylic acid to confer basal thermotolerance in *Arabidopsis thaliana*. *New Phytologist, 182*, 175–187.
Dar, T. A., Uddin, M., Khan, M. M. A., Hakeem, K. R., & Jaleel, H. (2015). Jasmonates counter plant stress: A review. *Environmental and Experimental Botany, 115*, 49–57.
Ding, C. K., Wang, C. Y., Gross, K. C., & Smith, D. L. (2001). Reduction of chilling injury and transcript accumulation of heat shock proteins in tomato fruit by methyl jasmonate and methyl salicylate. *Plant Science, 161*(6), 1153–1159.
Divekar, P. A., Narayana, S., Divekar, B. A., Kumar, R., Gadratagi, B. G., Ray, A., & Behera, T. K. (2022). Plant secondary metabolites as defense tools against herbivores for sustainable crop protection. *International Journal of Molecular Sciences, 23*(5), 2690.
Farmer, E. E., & Ryan, C. A. (1990). Interplant communication: Airborne methyl jasmonate induces synthesis of proteinase inhibitors in plant leaves. *Proceedings of the National Academy of Sciences, 87*(19), 7713–7716.
Farooq, M., Basra, S. M. A., Wahid, A., Ahmad, N., & Saleem, B. A. (2009). Improving the drought tolerance in rice (*Oryza sativa* L.) by exogenous application of salicylic acid. *Journal of Agronomy and Crop Science, 195*(4), 237–246.
Gimenez, E., Salinas, M., & Manzano-Agugliaro, F. (2018). Worldwide research on plant defense against biotic stresses as an improvement for sustainable agriculture. *Sustainability, 10*(2), 391.
Gong, Z. (2021). Plant abiotic stress: New insights into the factors that activate and modulate plant responses. *Journal of Integrative Plant Biology, 63*(3), 429.
Han, G. Z. (2017). Evolution of jasmonate biosynthesis and signaling mechanisms. *Journal of Experimental Botany, 68*(6), 1323–1331.
Hu, Y., Jiang, L., Wang, F., & Yu, D. (2013). Jasmonate regulates the inducer of cbf expression-c-repeat binding factor/dre binding factor1 cascade and freezing tolerance in *Arabidopsis*. *Plant Cell, 25*, 2907–2924.
Huan, C., Yang, X., Wang, L., Kebbeh, M., Wang, Y., Dai, B., & Zhou, H. (2022). Methyl jasmonate treatment regulates α-linolenic acid metabolism and jasmonate acid signaling pathway to improve chilling tolerance in both stony hard and melting flesh peaches. *Postharvest Biology and Technology, 190*, 111960.
Jagadish, S. V. K., Muthurajan, R., Oane, R., Wheeler, T. R., Heuer, S., Bennett, J., & Craufurd, P. (2010). Physiological and proteomic approaches to address heat tolerance during anthesis in rice (*Oryza sativa* L.). *Journal of Experimental Botany, 61*(1), 143–156.
Jiang, M., Xu, F., Peng, M., Huang, F., & Meng, F. (2016). Methyl jasmonate regulated diploid and tetraploid black locust (*Robinia pseudoacacia* L.) tolerance to salt stress. *Acta Physiologiae Plantarum, 38*, 1–13.
Jing, Y., Liu, J., Liu, P., Ming, D., & Sun, J. (2019). Overexpression of TaJAZ1 increases powdery mildew resistance through promoting reactive oxygen species accumulation in bread wheat. *Scientific Reports, 9*(1), 5691.
Kadyrzhanova, D. K., Vlachonasios, K. E., Ververidis, P., & Dilley, D. R. (1998). Molecular cloning of a novel heat induced/chilling tolerance related cDNA in tomato fruit by use of mRNA differential display. *Plant Molecular Biology, 36*, 885–895.

Koch, T., Bandemer, K., & Boland, W. (1997). Biosynthesis of cis-jasmone: A pathway for the inactivation and disposal of the plant stress hormone jasmonic acid to the gas phase? *Helvetica Chimica Acta, 80*(3), 838–850.

Kolaksazov, M., Laporte, F., Ananieva, K., Dobrev, P., Herzog, M., & Ananiev, E. D. (2013). Effect of chilling and freezing stresses on jasmonate content in *Arabis alpina*. *Bulgarian Journal of Agricultural Science, 19*(2), 15–17.

Koo, Y. J., Yoon, E. S., Seo, J. S., Kim, J. K., & Choi, Y. D. (2013). Characterization of a methyl jasmonate specific esterase in *Arabidopsis*. *Journal of the Korean Society for Applied Biological Chemistry, 56*, 27–33.

Lee, T. M., Lur, H. S., Lin, V. H., & Chu, C. (1996). Physiological and biochemical changes related to methyl jasmonate-induced chilling tolerance of rice (*Oryza sativa* L.) seedlings. *Plant, Cell & Environment, 19*, 65–74.

Li, C., Schilmiller, A. L., Liu, G., Lee, G. I., Jayanty, S., Sageman, C., & Howe, G. A. (2005). Role of β-oxidation in jasmonate biosynthesis and systemic wound signaling in tomato. *The Plant Cell, 17*(3), 971–986.

Li, R., Jin, J., Xu, J., Wang, L., Li, J., Lou, Y., & Baldwin, I. T. (2021). Long non-coding RNAs associate with jasmonate-mediated plant defence against herbivores. *Plant, Cell & Environment, 44*(3), 982–994.

Li, Y., Ye, W., Wang, M., & Yan, X. (2009). Climate change and drought: A risk assessment of crop-yield impacts. *Climate Research, 39*(1), 31–46.

Liu, D., Zhao, Q., Cui, X., Chen, R., Li, X., Qiu, B., & Ge, F. (2019). A transcriptome analysis uncovers *Panax notoginseng* resistance to *Fusarium solani* induced by methyl jasmonate. *Genes & Genomics, 41*, 1383–1396.

Lymperopoulos, P., Msanne, J., & Rabara, R. (2018). Phytochrome and phytohormones: Working in tandem for plant growth and development. *Frontiers in Plant Science, 9*, 1037.

Maksymiec, W., & Krupa, Z. (2007). Effects of methyl jasmonate and excess copper on root and leaf growth. *Biologia Plantarum, 51*, 322–326.

Mishra, V., Mishra, R., & Sharma, R. S. (2022). Ribosome-inactivating proteins: An unfathomed biomolecule for developing multi-stress tolerant transgenic plants. *International Journal of Biological Macromolecules, 210*, 107–122.

Moosa, A., Sahi, S. T., Khan, S. A., & Malik, A. U. (2019). Salicylic acid and jasmonic acid can suppress green and blue moulds of citrus fruit and induce the activity of polyphenol oxidase and peroxidase. *Folia Horticulturae, 31*(1), 195–204.

Motallebi, P., Niknam, V., Ebrahimzadeh, H., Hashemi, M., & Enferadi, S. T. (2017). Exogenous methyl jasmonate treatment induces defense response against *Fusarium culmorum* in wheat seedlings. *Journal of Plant Growth Regulation, 36*, 71–82.

Ndiaye, A., Diallo, A. O., Fall, N. C., Diouf, R. D., Diouf, D., & Kane, N. A. (2022). Transcriptomic analysis of methyl jasmonate treatment reveals gene networks involved in drought tolerance in pearl millet. *Scientific Reports, 12*(1), 5158.

Nguyen, D., Rieu, I., Mariani, C., & van Dam, N. M. (2016). How plants handle multiple stresses: Hormonal interactions underlying responses to abiotic stress and insect herbivory. *Plant Molecular Biology, 91*, 727–740.

Norastehnia, A., Sajedi, R. H., & Nojavan-Asghari, M. (2007). Inhibitory effects of methyl jasmonate on seed germination in maize (*Zea mays* L.): Effect on amylase activity and ethylene production. *General and Applied Plant Physiology, 33*, 13–23.

Onohata, T., & Gomi, K. (2020). Overexpression of jasmonate-responsive OsbHLH034 in rice results in the induction of bacterial blight resistance via an increase in lignin biosynthesis. *Plant Cell Reports, 39*, 1175–1184.

Park, H. J., Kim, W. Y., & Yun, D. J. (2013). A role for GIGANTEA: Keeping the balance between flowering and salinity stress tolerance. *Plant Signaling & Behavior, 8*, e24820.

Poonam, S., Kaur, H., & Geetika, S. (2013). Effect of jasmonic acid on photosynthetic pigments and stress markers in *Cajanus cajan* (L.) Millsp. seedlings under copper stress. *The Egyptian Journal of Botany, 53*(1), 179–187.

Qi, X., Chen, M., Liang, D., Xu, Q., Zhou, F., & Chen, X. (2020). Jasmonic acid, ethylene and ROS are involved in the response of cucumber (*Cucumis sativus* L.) to aphid infestation. *Scientia Horticulturae, 269*, 109421.

Qu, A. L., Ding, Y. F., Jiang, Q., & Zhu, C. (2013). Molecular mechanisms of the plant heat stress response. *Biochemical and Biophysical Research Communications, 432*(2), 203–207.

Quan, R., Lin, H., Mendoza, I., Zhang, Y., Cao, W., Yang, Y., & Guo, Y. (2007). SCABP8/CBL10, a putative calcium sensor, interacts with the protein kinase SOS_2 to protect Arabidopsis shoots from salt stress. *The Plant Cell, 19*(4), 1415–1431.

Ramanjulu, S., & Bartels, D. (2002). Drought-and desiccation-induced modulation of gene expression in plants. *Plant, Cell & Environment, 25*(2), 141–151.

Raza, A., Charagh, S., Najafi-Kakavand, S., & Siddiqui, M. H. (2021). The crucial role of jasmonates in enhancing heavy metals tolerance in plants. In *Jasmonates and Salicylates Signaling in Plants* (pp. 159–183). Wiley-Blackwell.

Raza, A., Mehmood, S. S., Tabassum, J., & Batool, R. (2019). Targeting plant hormones to develop abiotic stress resistance in wheat. In *Wheat Production in Changing Environments: Responses, Adaptation and Tolerance* (pp. 557–577). Wiley-Blackwell.

Roche, J., Hewezi, T., Bouniols, A., & Gentzbittel, L. (2007). Transcriptional profiles of primary metabolism and signal transduction-related genes in response to water stress in field-grown sunflower genotypes using a thematic cDNA microarray. *Planta, 226*(3), 601–617.

Rojo, E., Solano, R., & Sánchez-Serrano, J. J. (2003). Interactions between signaling compounds involved in plant defense. *Journal of Plant Growth Regulation, 22*, 82–98.

Santino, A., Taurino, M., De Domenico, S., Bonsegna, S., Poltronieri, P., Pastor, V., & Flors, V. (2013). Jasmonate signaling in plant development and defense response to multiple (a) biotic stresses. *Plant Cell Reports, 32*, 1085–1098.

Scalschi, L., Sanmartín, M., Camañes, G., Troncho, P., Sánchez-Serrano, J. J., García-Agustín, P., & Vicedo, B. (2015). Silencing of OPR3 in tomato reveals the role of OPDA in callose deposition during the activation of defense responses against *Botrytis cinerea. The Plant Journal, 81*(2), 304–315.

Schaller, A., & Stintzi, A. (2009). Enzymes in jasmonate biosynthesis: Structure, function, regulation. *Phytochemistry, 70*(13, 14), 1532–1538.

Schweizer, F., Bodenhausen, N., Lassueur, S., Masclaux, F. G., & Reymond, P. (2013). Differential contribution of transcription factors to *Arabidopsis thaliana* defense against *Spodoptera littoralis. Frontiers in Plant Science, 4*, 13.

Seng, S., Wu, C., Wu, J., Zhong, X., He, J., & Yi, M. (2020). Silencing GhCOI1 in *Gladiolus hybridus* increases susceptibility to *Alternaria brassicicola* and impairs inducible defenses. *Plant Cell, Tissue and Organ Culture (PCTOC), 140*, 69–81.

Sharma, M., & Laxmi, A. (2016). Jasmonates: Emerging players in controlling temperature stress tolerance. *Frontiers in Plant Science, 6*, 1129.
Shinozaki, K., Yamaguchi-Shinozaki, K., & Seki, M. (2003). Regulatory network of gene expression in the drought and cold stress responses. *Current Opinion in Plant Biology, 6*(5), 410–417.
Su, Y., Huang, Y., Dong, X., Wang, R., Tang, M., Cai, J., & Nie, G. (2021). Exogenous methyl jasmonate improves heat tolerance of perennial ryegrass through alteration of osmotic adjustment, antioxidant defense, and expression of jasmonic acid-responsive genes. *Frontiers in Plant Science, 12*, 664519.
Tan, C. W., Chiang, S. Y., Ravuiwasa, K. T., Yadav, J., & Hwang, S. Y. (2012). Jasmonate-induced defenses in tomato against *Helicoverpa armigera* depend in part on nutrient availability, but artificial induction via methyl jasmonate does not. *Arthropod-Plant Interactions, 6*, 531–541.
Tchounwou, P. B., Yedjou, C. G., Patlolla, A. K., & Sutton, D. J. (2012). Heavy metal toxicity and the environment. In *Molecular, Clinical and Environmental Toxicology: Volume 3: Environmental Toxicology* (pp. 133–164). Springer.
Ueda, J., & Saniewski, J. (2006). Methyl jasmonate-induced stimulation of chlorophyll formation in the basal part of tulip bulbs kept under natural light conditions. *Journal of Fruit and Ornamental Plant Research, 14*, 199–210.
Van Zelm, E., Zhang, Y., & Testerink, C. (2020). Salt tolerance mechanisms of plants. *Annual Review of Plant Biology, 71*, 403–433.
Verma, V., Ravindran, P., & Kumar, P. P. (2016). Plant hormone-mediated regulation of stress responses. *BMC Plant Biology, 16*, 1–10.
Walia, H., Wilson, C., Condamine, P., Liu, X., Ismail, A. M., & Close, T. J. (2007). Large-scale expression profiling and physiological characterization of jasmonic acid-mediated adaptation of barley to salinity stress. *Plant, Cell & Environment, 30*, 410–421.
Wang, Z., Tan, X., Zhang, Z., Gu, S., Li, G., & Shi, H. (2012). Defense to *Sclerotinia sclerotiorum* in oilseed rape is associated with the sequential activations of salicylic acid signaling and jasmonic acid signaling. *Plant Science, 184*, 75–82.
Wasternack, C. (2007). Jasmonates: An update on biosynthesis, signal transduction, and action in plant stress response, growth, and development. *Annals of Botany, 100*(4), 681–697.
Wasternack, C., & Feussner, I. (2018). The oxylipin pathways: Biochemistry and function. *Annual Review of Plant Biology, 69*, 363–386.
Wasternack, C., & Hause, B. (2013). Jasmonates: Biosynthesis, perception, signal transduction, and action in plant stress response, growth, and development. An update to the 2007 review in *Annals of Botany*. *Annals of Botany, 111*(6), 1021–1058.
Wasternack, C., & Kombrink, E. (2010). Jasmonates: Structural requirements for lipid-derived signals active in plant stress responses and development. *ACS Chemical Biology, 5*(1), 63–77.
Wasternack, C., & Song, S. (2017). Jasmonates: Biosynthesis, metabolism, and signaling by proteins activating and repressing transcription. *Journal of Experimental Botany, 68*(6), 1303–1321.
Weber, H., Vick, B. A., & Farmer, E. E. (1997). Dinor-oxo-phytodienoic acid: A new hexadecanoid signal in the jasmonate family. *Proceedings of the National Academy of Sciences of the USA, 94*, 10473–10478.

Woldemariam, M. G., Onkokesung, N., Baldwin, I. T., & Galis, I. (2012). Jasmonoyl-l-isoleucine hydrolase 1 (JIH1) regulates jasmonoyl-l-isoleucine levels and attenuates plant defenses against herbivores. *The Plant Journal, 72*(5), 758–767.

Xin, Z., Zhang, J., Ge, L., Lei, S., Han, J., Zhang, X., & Sun, X. (2017). A putative 12-oxophytodienoate reductase gene CsOPR3 from *Camellia sinensis* is involved in wound and herbivore infestation responses. *Gene, 615*, 18–24.

Xu, J., Wang, X., Zu, H., Zeng, X., Baldwin, I. T., Lou, Y., & Li, R. (2021). Molecular dissection of rice phytohormone signaling involved in resistance to a piercing-sucking herbivore. *New Phytologist, 230*(4), 1639–1652.

Yamaguchi-Shinozaki, K., & Shinozaki, K. (2006). Transcriptional regulatory networks in cellular responses and tolerance to dehydration and cold stresses. *Annual Review of Plant Biology, 57*, 781–803.

Yan, C., Fan, M., Yang, M., Zhao, J., Zhang, W., Su, Y., & Xie, D. (2018). Injury activates Ca2+/calmodulin-dependent phosphorylation of JAV1-JAZ8-WRKY51 complex for jasmonate biosynthesis. *Molecular Cell, 70*(1), 136–149.

Yan, J., Tan, Y., Lv, Y., Wang, F., Zhang, Y., & Chi, D. (2021). Effect of jasmonate treatments on leaves of *Rosa rugosa* 'Plena' and detoxification enzymes and feeding of adult *Monolepta hieroglyphica. Journal of Forestry Research, 32*(3), 1253–1261.

Yan, L., Zhai, Q., Wei, J., Li, S., Wang, B., Huang, T., & Li, C. (2013). Role of tomato lipoxygenase D in wound-induced jasmonate biosynthesis and plant immunity to insect herbivores. *PLoS Genetics, 9*(12), e1003964.

Yan, Y., Borrego, E., & Kolomiets, M. V. (2013). Jasmonate biosynthesis, perception, and function in plant development and stress responses. *Lipid Metabolism*. InTech, 393–442.

Yan, Z., Chen, J., & Li, X. (2013). Methyl jasmonate as a modulator of Cd toxicity in *Capsicum frutescens* var. fasciculatum seedlings. *Ecotoxicology and Environmental Safety, 98*, 203–209.

Yan, Z., Zhang, W., Chen, J., & Li, X. (2015). Methyl jasmonate alleviates cadmium toxicity in *Solanum nigrum* by regulating metal uptake and antioxidative capacity. *Biologia Plantarum, 59*(2), 373–381.

Ye, M., Luo, S. M., Xie, J. F., Li, Y. F., Xu, T., Liu, Y., & Zeng, R. S. (2012). Silencing COI1 in rice increases susceptibility to chewing insects and impairs inducible defense. *PLoS One, 7*(4), e36214.

Yuan, H. M., Liu, W. C., & Lu, Y. T. (2017). CATALASE2 coordinates SA-mediated repression of both auxin accumulation and JA biosynthesis in plant defenses. *Cell Host & Microbe, 21*(2), 143–155.

Zeng, J., Zhang, T., Huangfu, J., Li, R., & Lou, Y. (2021). Both allene oxide synthase genes are involved in the biosynthesis of herbivore-induced jasmonic acid and herbivore resistance in rice. *Plants, 10*(3), 442.

Zhang, B., Chen, K., Bowen, J., Allan, A., Espley, R., Karunairetnam, S., & Ferguson, I. (2006). Differential expression within the LOX gene family in ripening kiwifruit. *Journal of Experimental Botany, 57*(14), 3825–3836.

Zhang, H., Zhang, Q., Zhai, H., Li, Y., Wang, X., Liu, Q., & He, S. (2017). Transcript profile analysis reveals important roles of jasmonic acid signaling pathway in the response of sweet potato to salt stress. *Scientific Reports, 7*(1), 40819.

Zhang, L., Zhang, F., Melotto, M., Yao, J., & He, S. Y. (2017). Jasmonate signaling and manipulation by pathogens and insects. *Journal of Experimental Botany, 68*(6), 1371–1385.

Zhang, P. J., Zhao, C., Ye, Z. H., & Yu, X. P. (2020). Trade-off between defense priming by herbivore-induced plant volatiles and constitutive defense in tomato. *Pest Management Science, 76*(5), 1893–1901.

Zhang, Y. T., Zhang, Y. L., Chen, S. X., Yin, G. H., Yang, Z. Z., Lee, S., & Yang, F. S. (2015). Proteomics of methyl jasmonate-induced defense response in maize leaves against Asian corn borer. *BMC Genomics, 16*(1), 1–16.

Zhao, Q., Sun, Q., Dong, P., Ma, C., Sun, H., & Liu, C. (2019). Jasmonic acid alleviates boron toxicity in *Puccinellia tenuiflora*, a promising species for boron phytoremediation. *Plant and Soil, 445*, 397–407.

Zhao, S., & Li, Y. (2021). Current understanding of the interplays between host hormones and plant viral infections. *PLoS Pathogens, 17*(2), e1009242.

Zhao, S., Zhang, Q., Liu, M., Zhou, H., Ma, C., & Wang, P. (2021). Regulation of plant responses to salt stress. *International Journal of Molecular Sciences, 22*(9), 4609.

Zhu, J. K. (2002). Salt and drought stress signal transduction in plants. *Annual Review of Plant Biology, 53*(1), 247–273.

CHAPTER 5

Interplay of Methyl Jasmonate with Other Phytohormones During Abiotic Stress Tolerance: An Overview

BILAL AHMAD,[1,2] TARIQ AHMAD DAR,[1,3] URWAT UL MUSKAN,[4] SYED HUMAIRA,[4] AJMAT JAHAN,[1] MUDASIR IRFAN DAR,[2] and SUHAIL BASHIR[2]

[1]*Advanced Plant Physiology Section, Department of Botany, Aligarh Muslim University, Aligarh, Uttar Pradesh, India*

[2]*Higher Education Department Jammu and Kashmir India*

[3]*Sri Pratap College Cluster University Srinagar*

[4]*Department of Botany, University of Kashmir, Jammu and Kashmir, India*

ABSTRACT

Plants keep a continuous watch not only on the extrinsic environment but also their inherent developmental status so as to fine-tune their growth and differentiation under normal as well as perturbed environments with the help of a highly complex and organized regulatory system comprising of an array of regulatory proteins and signaling molecules. Among the diverse regulatory signals, phytohormones constitute the most imperative class of signaling compounds that operate through diverse routes thereby providing plasticity to acclimatize to the persistently changing developmental as well as environmental cues. Extensive studies at the molecular level have identified explicit, non-redundant and overlapping signal transduction pathway for each hormone explaining the overlap in

Biotechnological Applications of Jasmonates: A Group of Less Well-Known Plant Growth Regulators. Tariq Ahmad Dar, Bilal Ahmad, and Khalid Rehman Hakeem (Eds.)

the phytohormone functions which in-turn signifies the essence of phytohormone molecular cross-talk studies in gaining insights regarding the hormonal responses in different plants. Jasmonic acid (JA) being a vital plant stress regulator plays a pivotal role during various stress responses of both biotic as well as abiotic type in addition to the normal growth and developmental aspects. Molecular studies during the last two decades led to a noteworthy progress in gaining understanding about the biosynthesis and signaling of JA, and its cross-talk with other stress regulators particularly under environmental perturbations. This review is an attempt to gain insights into the interplay of JA with different phytohormones operational during abiotic stress mitigation in plants highlighting the recent progress achieved in elucidating the role of vital regulatory proteins at the transcription as well as translational levels.

5.1 INTRODUCTION

Plants encounter a diverse range of perturbations in natural environments both of biotic and abiotic nature and concurrent exposure to stressors of different nature leads to the alteration of a wide range of physio-morphological traits in them, thereby exacerbating the capacity to endure. Therefore, plants have developed miscellaneous strategies to counter the multifactorial environmental perturbations and a variety of signals operate in a coordinated manner allowing the plants to survive even under extreme stresses (Shalak et al., 2021). Plants not only scrutinize their intrinsic developmental status and exterior environment constantly but also develop adjustments in order to establish a balance with regard to growth and differentiation under normal as well as perturbed environments with the help of an exceedingly complex, interconnected and organized regulatory network comprising of diverse types of regulatory proteins and signaling molecules (Huang et al., 2022; Jang et al., 2020; Durbak et al., 2012). Phytohormones constitute an imperative class of signaling molecules which operate through diverse channels conferring plasticity to constantly transforming developmental as well as environmental cues in addition to helping the plants to adapt to the same (Aftab and Roychoudhury, 2021; Ali et al., 2020; Per et al., 2018; McSteen and Zhao, 2008). Plants, in comparison to animals, can unremittingly cease and restart growth due to the plasticity in their patterns of growth and architecture which is partly accomplished by certain endogenously synthesized substances diverse in structure and efficient in minute concentrations named as phytohormones (Durbak et al., 2012; McSteen and Zhao, 2008). Several plant hormone types have been identified until recently and extensive and thorough studies have revealed that each hormone involves an explicit, non-redundant signaling pathway; nevertheless, the functions of different plant hormones chiefly overlap in various plant and growth development aspects as well as diverse environmental responses, signifying the importance of phyto-hormone signaling cross-talk studies in establishing valuable insights into different responses of plants to various environmental constraints (Aftab and Roychoudhury, 2021; Ku et al., 2018; Durbak et al., 2012; McSteen and Zhao, 2008). However, the most difficult challenge of the present time is to gain an inclusive understanding about the interac-tive mechanism and coordination between genes and phytohormones throughout the different stages of plant development particularly under challenging circumstances (different abiotic and biotic stresses) (Aftab and Roychoudhury, 2021; Ku et al., 2018). The adaptive as well as tolerance response that becomes operational in plants during stressful conditions chiefly depends on the chemical signals (secondary metabo-lism products) which are predominantly orchestrated by phytohormones by upholding a multifaceted balance to sustain the growth as well as tolerance (Sreenivasulu et al., 2012; Durbak et al., 2012; McSteen and Zhao, 2008). It is well recognized and previously established fact that the phytohormones are not operative discretely, instead manipulate each other at various steps (biosynthesis or signaling) in order to organize different developmental and environmental signaling pathways (Delgado et al., 2021; Ku et al., 2018; Gray, 2004). The signaling pathways of phytohormones are interconnected through synergistic or antagonistic cross-talk and these alter the biosynthesis and/or responses of each other (Kazan, 2015). Hormonal interplay can markedly determine the biosynthesis and signal transduction of various hormones and may involve the feedback of both positive as well as negative type (Hu et al., 2017; Robert-Seilaniantz et al., 2007). However, it is indispensable to be acquainted with such a vast majority of response mechanisms and the signaling cross-talk cascades allowing the feedback shift among different signaling pathways (Ku et al., 2018). The signaling pathways of different phytohormones are latticed in a complicated network imparting the plants a massive regulatory power to swiftly adapt to the variable environment facilitating them to efficiently exploit the limited resources which otherwise can affect the growth, development as well as adaptation negatively (Pieterse et al., 2012). Owing to the raising significance of cross-talk studies of diverse exogenous and endogenous growth regulators and other signaling molecules in exemplifying the role in resistance as well as adaptation against environmental stresses, continuous efforts are underway to gain a comparatively inclusive understanding regarding the different response pathways and the points where these pathways switch. Among the different signaling regulatory molecules, calcium sensors constitute the initial line of response against exterior stimuli which thereafter triggers the downstream signal transduction. Abscisic acid is regarded as stress-regulating phytohormone, and it works together-with the salicylic acid and jasmonic acid signaling pathways in order to channel different resources concerning the alleviation of stresses of diverse nature (Aftab and Roychoudhury, 2021; Ku et al., 2018; Verma et al., 2016). G-proteins also known as GTP-binding proteins constitute a structurally and functionally diverse class of proteins primarily involved in transduction of these signals in different hormonal signaling pathways (Aftab and Roychoudhury, 2021).

One among the vital hormones acting as stress modulators, jasmonic acid, occupies an indispensable position as a chemical signal responsible for the regulation of growth and development chiefly under perturbed environments (Huang et al., 2022; Jang et al., 2020; Yang et al., 2020). JAs, a very important group of elicitors in plants, participate in a wide array of plant metabolic pathways; do not act discretely but in a complex relationship with other phytohormones involving signaling cross-talk with them (Per et al., 2018; Ahmad et al., 2016; Wasternack and Strnad, 2016). MeJA is one among the fundamental cellular regulator which arbitrates an array of developmental and defense responses against diverse perturbations of biotic as well as abiotic nature. MeJA, being pleiotropic, is intangible with regard to its biogenesis, regulation, signaling cross-talk and mode of action (Huang et al., 2022; Jang et al., 2020). One of the imperative break-through which helped in gaining insights into the role of MeJA in gene up-regulation as well as long distance systemic signaling evolved with the characterization of a vital gene encoding jasmonic acid carboxyl methyl-transferase (Huang et al., 2022; Jang et al., 2020; Yang et al., 2020). Functional genomics and bioinformatic studies of jasmonates have helped in identification of a set of MeJA responsive genes, as well as the use of diverse volatile signals by plants which help them endure miscellaneous and unpredictable environments. Comprehensive studies have revealed an intensive cross-talk (synergistic and/or antagonistic), between the signal transduction pathways of various other phytohormones and JAs to achieve the purpose of stress tolerance, both biotic as well as abiotic (Ahmad et al., 2016; Wasternack and Strnad, 2016; Kazan and Manners, 2012). This cross-talk among various plant hormones has the capability of reprogramming the genetic machinery and reducing the impacts of stress, thus leading to the improved stress tolerance in plants (Aftab and Roychoudhury, 2021; Per et al., 2018). Like many other phenomena, fertility and floral development are coordinately modulated by various endogenously produced developmental signals including hormones (auxins, gibberellins, and jasmonates), environmental cues, etc. (Per et al., 2018; Durbak et al., 2012; McSteen and Zhao, 2008). The intricate dialogue of phytohormones is a decisive component of plant metabolism determining growth and development both under normal and extreme environments. The interactive studies of JAs reveal that such interplay, agonistic as well as antagonistic, is imperative in order to regulate various growth and developmental aspects. AUXs, CKs, GAs, and BRs, usually display an antagonistic type of interaction with JAs during plant development though not always. For-example; GAs and JAs display an antagonist interaction during stem elongation, but a synergistic one in stamen development; likewise, an antagonistic interplay with auxin to regulate apical root growth, but a synergistic one to promote lateral root growth (Per et al., 2018; Cai et al., 2014; Qi et al., 2014; Ren et al., 2009). JA interacts with SA to modulate plant defenses biotic as well as abiotic stressors. Likewise, interaction with ABA helps in the regulation of cellular metabolism while interplay of JAZs with PYRABACTIN RESISTANCE 1-Like proteins (ABA receptor) modulates JA signaling (Per et al., 2018; Cai et al., 2014; Ren et al., 2009). JAs have also been observed to regulate androecium development and in-turn fertility, development of trichomes and root growth under stressful conditions, and diverse defense responses against biotic stressful agents like insects and herbivores (Huang et al., 2022). In the following section, an attempt is made to briefly cover the cross-talk mechanism of JA with some selective phytohormones during abiotic stress.

5.2 INTERPLAY OF JASMONATES WITH OTHER PHYTOHORMONES

5.2.1 JASMONATES AND AUXIN

Various studies have accumulated enough evidence regarding the existence of association between auxin and JAs signaling pathways. Indole-3-acetic acid has been observed to be capable of stimulation of the expression of *JAZ1/TIFY10A*, a JA signaling repressor of Arabidopsis (Grunewald et al., 2009). JA down-regulates and suppresses the expression of PLETHORA (PLT1 and PLT2) TFs which are the chief regulators of auxin-induced root-meristem development thereby inhibiting auxin-mediated root growth in plants (Chen et al., 2011). It has been documented that the JA-mediated PLT expression and regulation of root-stem cell niche was through the involvement of *MYC2*, a master regulator of the signal transduction pathway of JA (Chen et al., 2011). Moreover, the investigations of Tiryaki and Staswick (2002) evolved a dual conclusion that the auxin and JA-induced signaling can act both synergistically or antagonistically. The authors observed that the exogenous auxin application either induced or repressed the JA-responsive genes. The biological context as well as the underlying mechanism is yet to be understood, however, one of the possible reasons suggested is the acyl acid amido synthetases, belonging to the GH3 family, which are responsible for the IAA and JA amino acid conjugation. An active signal is generated by this amino acid conjugation in case of JA, but the case is different for IAA where such conjugation might lead to the inactivation (Khan and Stone, 2007; Staswick and Tiryaki, 2004). Furthermore, during the lateral root formation induced by JA which involves an ethylene response factor 109 (ERF 109), cross-talk between auxin and JA has been suggested Cai et al. (2014). Moreover, WRKY57 in *Arabidopsis*, has been observed to function as the convergence node in IAA and JA-mediated signaling during the leaf senescence induced by JA (Jiang et al., 2014). Studies in *Arabidopsis* have revealed that the JA and IAA signaling repressors (*JAZ4/8* and *IAA29*, respectively) both bind *WRKY57* competitively, but IAA up-regulates it while JA results in its down-regulation. Additionally, studies on the signal cross-talk of three different phytohormones (auxin, JA and ABA) revealed that these can engage the TOPLESS co-repressor proteins by interacting with certain explicit adaptor proteins (Pauwels et al., 2010). The adaptor proteins for the three hormones auxin, JA and ABA have been found to be AUX/IAA, NINJA, and AFP, respectively (Pauwels et al., 2010). Such valuable investigations have developed a proper understanding regarding the molecular mechanisms involved during the stress-related signal transduction cascades so as to regulate expression of gene in plants (Pauwels et al., 2010).

5.2.2 JASMONATES AND CYTOKININS

Evidence regarding the signaling cross-talk of JA and CK is scarce; however, an antagonistic link between the two hormonal pathways can be established (Stoynova-Bakalova et al., 2008; Naik et al., 2002). Exogenous sourcing of CK as well as CK *in-vivo* is known to speed-up the stress response mediated by JA (Dervinis et al., 2010; Sano et al., 1996) whereas application of JA is evidenced to induce CK ribosides' accumulation (Dermastia et al., 1994). An optimum level of CK concentration is always necessary for the proper growth of plants and can be maintained by exogenous application of MeJA, which is also known to modulate the activity of cytokinin oxidase and its expression (Avalbaev et al., 2016). Salinity stress exposed wheat seedlings when sourced with exogenous MeJA are known to regulate the cytokinin content by modulating the cytokinin oxidase activity (Avalbaev et al., 2016). Salt stress may lead to the conditions of water scarcity (drought). Biosynthesis of JA is speeded-up in the root tissues during drought conditions (Poltronieri et al., 2013). Biosynthesis and signaling of CK is repressed under such conditions which encourages the root system expansion, thereby acting on stress tolerance positively (Werner et al., 2010). This leads to the conclusion that CK and JA biosynthesis/signaling may possibly act antagonistically. This view finds further support from the studies of Liu et al. (2015) on rice. The authors observed that the exogenous sourcing of CK and JA regulate the gene expression associated to leaf senescence in an antagonistic manner and this regulation is not only determined by the levels of CK and JA but also by a balance between the levels of the two (Figure 5.1).

5.2.3 JASMONATES AND ETHYLENE

Ethylene (ET) and JA have been studied to function synergistically or antagonistically during plant development as well as during different stress responses (Zhu and Lee, 2015; Zhu, 2014). Such a cross-talk between the signaling pathways of ET and JA in plants facing abiotic stress has been explained in recent times (Kazan, 2015). Both JA and ET are known to activate ethylene response factor 1 (AtERF1) expression in *Arabidopsis*

(Lorenzo et al., 2003). The report has recently come to light that synergistic stimulation of *AtERF1* in *Arabidopsis* is necessary for salinity and drought stress alleviation (Cheng et al., 2013). Moreover, the interplay between the components of signal transduction of JA and ET finds another link through ETHYLENE INSENSITIVE3 (*EIN3*), an ET-activated TF and EIN3-Like1 (EIL1) which is a close homolog of *EIN3* (Zhang et al., 2014; Song et al., 2014b; Chang et al., 2013). MYC2, a JA-activated TF, binds to the promoter (EIN3 BINDING F-BOX PROTEIN1) of an *F-box* gene, *EIN3* and promotes its degradation thereby inducing its expression. On the other hand, MYC2 interacts with EIN3 physically thereby inhibiting its DNA binding activity. Lack of function-mutants of *CBF2/CBF3, EIN2* or *ORE9* were observed to show delay in the leaf senescence promoted by JA, ET as well as ABA (Woo et al., 2013; Kim et al., 2011; Sharabi-Schwager et al., 2010). Thus, overlapping functions of phytohormones can be observed regulating the processes like leaf senescence via common genetic pathways. Thus, this sort of regulation by phytohormones appears to be a multifaceted process consistent with which is a webbed regulatory mechanism (Breeze et al., 2011). Lastly, while studying the salinity stress response in tobacco Zhang et al. (2004) observed that JA connects to ET as well as ABA signaling, through ERF (*JERF1*) which is a jasmonate responsive TF found in tomato. Additionally, another tomato ERF (*SlERF.B.3*) has been observed to be associated with cold and salt stress response (Klay et al., 2014).

5.2.4 JASMONATES AND GIBBERELLIC ACID

GAs may lead to either suppression or promotion of growth during explicit abiotic stress conditions (Colebrook et al., 2014). JAZ and DELLA proteins have been observed to interact directly with each other during the cross-talk between JA and GA. DELLA proteins, the chief repressors of the signal transduction pathway of GA, stimulate the competitive binding of JAZ with MYC2, a TF which activates JA responsive genes thereby modulating the JA signaling pathway (Hou et al., 2010). When JA is present, the fate of JAZ proteins is proteasome-mediated degradation. Likewise, GA when present induces proteasome-stimulated degradation of DELLA proteins. When GA is deficient, the DELLA proteins competitively bind to JAZs causing the release of MYC2, the alternate partner, which in turn attenuates the JA-responsive genes activating their expression (Hou et al., 2010; Boter et al., 2004). Thus, the MYC2 repression and JA signaling attenuation are completed. It is remarkable noting that C-termini of JAZs with JA-associated domain is very essential for interplay between JAZs and MYC2, and between JAZs and DELLAs as well (Hou et al., 2010; Chini et al., 2007). Fascinatingly, promoter of *RGL3*, a *DELLA* gene, has been shown to bind MYC2, leading to transcriptional upregulation by JA signaling (Wild et al., 2012). Additionally, the C-termini are reported to be necessary for the interaction between JAZs and COI1 and for binding of COR to different complexes (Katsir et al., 2008). Various other TFs in addition to MYC2 like MYC3, MYC4, MYB21, MYB24 are capable of binding and interaction with the DELLAs genes which has helped in gaining insights regarding the synergistic interaction between GA and JAs (Figure 5.1) (Lyons et al., 2013).

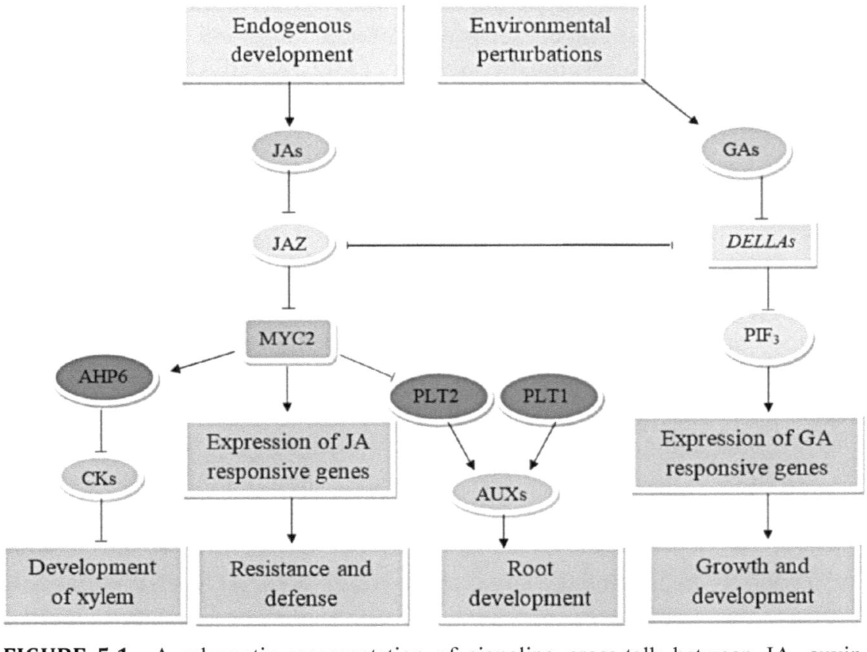

FIGURE 5.1 A schematic representation of signaling cross-talk between JA, auxin, cytokinin, and gibberellic acid. PLT1 and PLT2 act as key transcriptional regulators during root development. JA causes inhibition of root growth which is mediated through MYC2 by reduced expression of PLT1 and PLT2. JA and CK interaction in xylem development is antagonistic and the MYC2 mediates this process. MYC2 promotes the expression of AHP6 which acts as a cytokinin signaling inhibitor. The arrows indicate positive while T bars indicate negative regulation.

5.2.5 JASMONATES AND ABSCISIC ACID

Cross-talk amid signaling pathways of ABA and JA is not unexpected given their participation during various stress responses and may also involve MYC2 TF (Cutler et al., 2010). JA has originally been described as an ABA signaling activator (Abe et al., 2003). In addition to MYC2 and TOPLESS, cross-talk between ABA and JA might be facilitated by MAP kinase (MPK6) like proteins (Xing et al., 2008; Takahashi et al., 2007). One more protein AUXIN RESISTANT 1 which constitutes the sub-component of the RUB1-activating enzyme involved in the regulation of the Skp1–Cullin–F-box protein degradation activity (Tiryaki, 2007; Tiryaki and Staswick, 2002). Earlier, two mutants namely jasmonic acid insensitive4 (jin4) and jasmonic acid resistant1 (jar1) were reported to show an antagonistic interaction between ABA with JA interplay during the study of ABA-induced seed-germination inhibition (Staswick et al., 1992). This sort of antagonistic interplay amid JA and ABA signaling was further reported in *Oryza sativa* during the JA and ABA-mediated expression of salinity stress-inducible transcripts (Moons et al., 1997). During the JA signaling, regulation of metabolic reprogramming in Arabidopsis was observed to involve *NtPYL4, PYL4, ABA* receptors (Lackman et al., 2011). Furthermore, *NtPYL4* is a tobacco gene induced by JA but it known to encode a functional ABA receptor (Lackman et al., 2011). The detailed study by the authors led to the development of link between JA responses and the core ABA signaling pathway. Such an interaction between the two phytohormones has further been observed to regulate the responses of plants exposed to water scarcity (de Ollas et al., 2015). Over-expression of TaOPR1 encoding 12-oxophytodienoate reductase 1 (OPR1), a wheat gene in Arabidopsis led to salt stress mitigation. The *TaOPR1* overexpression resulted in enhanced ROS detoxification and the suggested mechanism was found to be dependent on ABA (Dong et al., 2013). Though, further study is needed to investigate the association of JAs with TaOPR1-mediated salt tolerance as TaOPR1 expression was not observed to have an immediate effect on JA signaling (Figure 5.2) (Dong et al., 2013).

5.2.6 JASMONATES AND SALICYLIC ACID

One of the most fascinating phytohormonal cross-talk has been witnessed during the signaling cascades of JA and salicylic acid (SA) (Khan et al.,

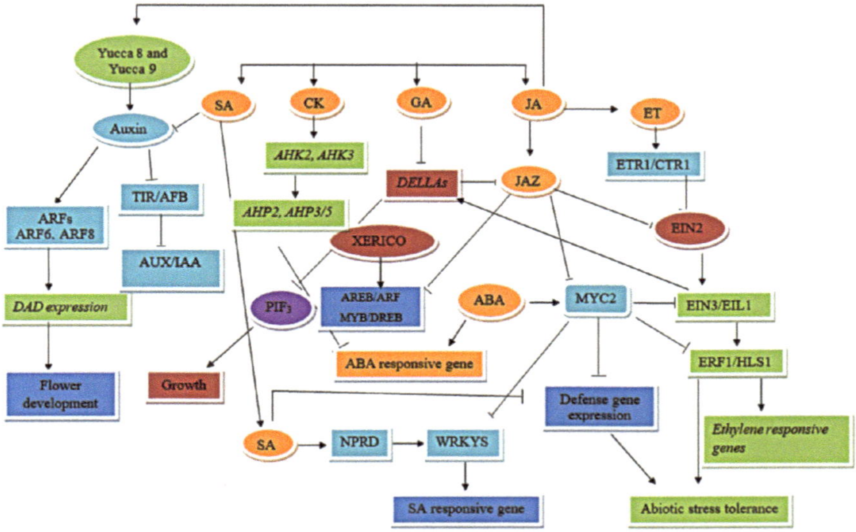

FIGURE 5.2 Schematic diagram highlighting signaling cross-talk of different hormones particularly that of jasmonic acid. Positive regulatory actions have been highlighted by arrows while negative regulatory actions are indicated by lines with bars. JA and GA interaction involves MYC2 as the main component. Interference of JAZ repressors limits the activity of DELLA on PIF3 promoting growth. DELLAs transduce the signal to JAZ repressors, resulting in the release of MYC2 from JAZ repression thereby contributing to JA-mediated defense response by stimulation of MYC2. On the other hand, reciprocal interaction between DELLAs and JAZs cause the release of downstream targets (e.g., MYC2 and PIF3) to regulate the JA and GA signaling antagonism. ABA, however, up-regulates MYC2. ET receptors (ETR1/CTR1) after perceiving the ET signals, the ET receptor complex is inactivated resulting in the stability of EIN3 and EIL1 which thereafter activate ET responses. EIL1/EIN3 are inhibited by JAZ, mediating synergy of JA and ET signal transduction in plant resistance, while the reciprocal interaction between EIN3/EIL1 and MYC2 mediates the ET and JA signaling antagonism. CK and JA regulate the production of each other under stressful conditions. Studies involving the YUCCA8 and YUCCA9 gene expression in Arabidopsis have revealed association of auxin homeostasis with the JA signal transduction pathway. In the Arabidopsis double mutant, the DAD1expression, an important JA biosynthetic gene, was markedly reduced, signifying the requirement of ARF6 and ARF8 for the stimulation of DAD1 expression. Redrawn with some modifications from Per et. al., (2018).

Abbreviations used: ARF: auxin response factor; ABA: abscissic acid; CTR1: CONSTITUTIVE TRIPLE RESPONSE 1; ETR1: ETHYLENE RESPONSE1; ERF1: ethylene response factor 1; EIL1: ethylene insensitive 3-like1 protein; ERF1: ethylene response factor 1; HLS1: HOOKLESS 1; JA: jasmonate; JAZ: jasmonate ZIM domain; EIN3: ethylene insensitive; SA: salicylic acid; NPRD: WRKYS (TF); GA: gibberellic acid; CK: cytokinin; DAD1: DEFECTIVE IN ANTHER DEHISCENCE1; and PIF3: PHYTOCHROME INTERACTING FACTOR 3;

2012b). Both JA and SA have been typically linked with the biotic stress, nevertheless, a collective approach of transcriptomics and proteomics has led to the proper identification of explicit proteins linked with varied abiotic and oxidative stress responses but their expression is regulated by SA and JA (Proietti et al., 2013). The prime approach of interaction between the two hormones is of mutual antagonism and a number of studies support such type of an interaction between the signaling pathways of the two (van der Does et al., 2013; Spoel et al., 2003). LOX2 and AOS, two important enzymes involved in the biosynthesis of JA, are soft targets of SA and hence suppressed by it (Laudert and Weiler, 1998; Spoel et al., 2003). In addition, SA can also target the JA biosynthesis at various points downstream (Leon-Reyes et al., 2010). Moreover, SA can suppress the AOS induction during wounding which also elucidates a negative interplay between the SA and JA signaling pathways (Harms et al., 1998). The antagonism between the signaling of the two hormones is also influenced by glutaredoxin (GRX480) (Meyer, 2008). GRX480 reduces thiol disulfide and is involved in redox regulation of proteins which contribute to a wide array of cellular phenomena (Meyer, 2008). SA-responsive PR gene regulation due to the interaction between GRX480 and TGA TFs has been established and NPR1 has been observed to be necessary for GRX480 expression via SA and TGA TFs (Ndamukong et al., 2007). GRX480 has also been observed to inhibit the expression of *PDF1.2*, a JA-responsive gene (Ndamukong et al., 2007). Interestingly, SA-stimulated NPR1 can lead to activation of GRX480 which can result in complex formation along with TGA factors leading to the repression of the JA-responsive gene expression. The positive role of SA in abiotic stresses of varied type (salt, heat, and osmotic stress) finds wide acceptance and has been comprehensively reviewed (Miura and Tada, 2014; Pal et al., 2013; Horváth et al., 2007). The point of convergence linking the *Arabidopsis* JA and SA signaling pathways has been elucidated to be MAP Kinase 4 (AtMPK4), which has the negative regulatory influence on SA activation (Brodersen et al., 2006). AtMPK4 represses the JA-mediated defense under conditions of biotic stress. Whether AtMPK4 plays the same role during abiotic stress needs further elucidation, but it is well established that AtMAPK4 is swiftly upregulated during stresses of abiotic nature (Ichimura et al., 2000). In this context, the role SA plays in the closure of stomatal is remarkable (Miura and Tada, 2014). Though synergistic cross-talk between SA and JA is also evidenced but reports to support are too scanty (Figure 5.2).

5.2.7 JASMONATES AND NITRIC OXIDE

Meager information regarding the JA and nitric oxide (NO) interaction is available. Few of JA-activated signaling reactions are linked with NO signaling. NO stimulates the biosynthesis of JAs in plants by regulating the expression of various genes involved in JA biosynthetic pathway like LOX3 (At1g17420), OPR 1, 2, 3 (Mur et al., 2013, 2012; Palmieri et al., 2008). However, NO has also been observed to repress an important JA biosynthetic pathway enzyme-AOC (Palmieri et al., 2008). The down-regulation of NO on AOC has been endorsed to S-nitrosylation (Romero-Puertas et al., 2008). Furthermore, treatment of the NG-nitro-L-arginine methyl ester (L-NAME), an NO synthase inhibitor modulates JA-induced closure of stomata signifying that NO operates downstream of JA (Xin et al., 2005).

5.3 CONCLUSION AND FUTURE PROSPECTS

Even though many studies have been conducted during the last two decades to gain an ample understanding about the JA signaling pathway and its interplay with other phytohormones; still, the currently available information about its role in various types of stresses is inadequate, primarily because of the composite and multifaceted cross-talk networks among multiple stresses and various signaling pathways. Even, the comprehensive mechanism of JA signaling pathway at the gene level remains elusive and the identified machinery in phytohormone signal transduction cascades are incomplete as many intervening signaling molecules and transcription factors are yet to be identified. Additionally, due to the presence of enormous cell membrane receptors and kinases, multiple stressors may lead to the activation of multiple enzymes, secondary messengers, and TFs, consequently resulting in the formation of extremely complex signaling web very difficult to elucidate. Currently, a lot of voids exist in comprehending the JA cross-talk particularly in the perception of multiple stress signals. However, omics studies may prove handy in elucidating the composite protein interface to identify the multifaceted stress signaling pathways and multifarious signaling proteins, messengers and TFs-mediated phytohormone cross-talks.

KEYWORDS

- **biogenesis**
- **calcium sensors**
- **membrane receptors**
- **phytohormones**
- **signaling molecules**
- **signaling pathway**

REFERENCES

Aftab, T. and Roychoudhury, A. (2021). Crosstalk among plant growth regulators and signaling molecules during biotic and abiotic stresses: Molecular responses and signaling pathways. *Plant Cell Reports*, 1–3.

Ahmad, P., Rasool, S., Gul, A., Akram, N. A., Ashraf, M., & Gucel, S. (2016). Jasmonates: multifunctional roles in stress tolerance. *Frontiers in Plant Science*, *7*, 813.

Ali, M. S. and Baek, K. H. (2020). Jasmonic acid signaling pathway in response to abiotic stresses in plants. *International Journal of Molecular Sciences*, *21*(2):621.

Avalbaev, A., Yuldashev, R., Fedorova, K., Somov, K., Vysotskaya, L., Allagulova, C., & Shakirova, F. (2016). Exogenous methyl jasmonate regulates cytokinin content by modulating cytokinin oxidase activity in wheat seedlings under salinity. *Journal of Plant Physiology*, *191*, 101–110.

Boter, M., Ruíz-Rivero, O., Abdeen, A., & Prat, S. (2004). Conserved MYC transcription factors play a key role in jasmonate signaling both in tomato and Arabidopsis. *Genes & Development*, *18*, 1577–1591.

Brodersen, P., Petersen, M., Bjørn Nielsen, H., Zhu, S., Newman, M. A., Shokat, K. M., et al. (2006). Arabidopsis MAP kinase 4 regulates salicylic acid- and jasmonic acid/ethylene-dependent responses via EDS1 and PAD4. *Plant Journal*, *47*, 532–546.

Browse, J. (2009). Jasmonate passes muster: A receptor and targets for the defense hormone. *Annual Review of Plant Biology*, *60*, 183–205.

Cai, X. T., Xu, P., Zhao, P. X., Liu, R., Yu, L. H., & Xiang, C. B. (2014). Arabidopsis ERF109 mediates cross-talk between jasmonic acid and auxin biosynthesis during lateral root formation. *Nature Communications*, *5*, 5833.

Chang, K. N., Zhong, S., Weirauch, M. T., Hon, G., Pelizzola, M., Li, H., Huang, S. S. C., Schmitz, R. J., Urich, M. A., Kuo, D., & Nery, J. R. (2013). Temporal transcriptional response to ethylene gas drives growth hormone cross-regulation in Arabidopsis. *Elife*, *2*.

Chen, Q., Sun, J., Zhai, Q., Zhou, W., Qi, L., Xu, L., Wang, B., Chen, R., Jiang, H., Qi, J., & Li, X. (2011). The basic helix-loop-helix transcription factor MYC2 directly represses

PLETHORA expression during jasmonate-mediated modulation of the root stem cell niche in Arabidopsis. *Plant Cell*, *23*, 3335–3352.

Cheng, Z., Sun, L., Qi, T., Zhang, B., Peng, W., Liu, Y., & Xie, D. (2011). The bHLH transcription factor MYC3 interacts with the jasmonate ZIM-domain proteins to mediate jasmonate response in Arabidopsis. *Molecular Plant*, *4*, 279–288.

Colebrook, E. H., Thomas, S. G., Phillips, A. L., & Hedden, P. (2014). The role of gibberellin signaling in plant responses to abiotic stress. *Journal of Experimental Biology*, *217*, 67–75.

Cutler, S. R., Rodriguez, P. L., Finkelstein, R. R., & Abrams, S. R. (2010). Abscisic acid: Emergence of a core signaling network. *Annual Review of Plant Biology*, *61*, 651–679.

de Ollas, C., Arbona, V., & Gómez-Cadenas, A. (2015). Jasmonoyl isoleucine accumulation is needed for abscisic acid build-up in roots of Arabidopsis under water stress conditions. *Plant Cell and Environment*, *38*, 2157–2170.

Delgado, C., Mora-Poblete, F., Ahmar, S., Chen, J. T., & Figueroa, C. R. (2021). Jasmonates and plant salt stress: Molecular players, physiological effects, and improving tolerance by using genome-associated tools. *International Journal of Molecular Sciences*, *22*(6), 3082.

Dermastia, M., Ravnikar, M., Vilhar, B., & Kovač, M. (1994). Increased level of cytokinin ribosides in jasmonic acid-treated potato (*Solanum tuberosum*) stem node cultures. *Physiologia Plantarum*, *92*, 241–246.

Dervinis, C., Frost, C. J., Lawrence, S. D., Novak, N. G., & Davis, J. M. (2010). Cytokinin primes plant responses to wounding and reduces insect performance. *Journal of Plant Growth Regulation*, *29*, 289–296.

Dong, W. et al. (2013). Wheat oxophytodienoate reductase gene TaOPR1 confers salinity tolerance via enhancement of abscisic acid signaling and reactive oxygen species scavenging. *Plant Physiology*, *161*, 1217–1228.

Durbak, A., Yao, H., & McSteen, P. (2012). Hormone signaling in plant development. *Current Opinion in Plant Biology*, *15*(1), 92–96.

Gray, W. M. (2004). Hormonal regulation of plant growth and development. *PLoS Biology*, *2*.

Grunewald, W., Vanholme, B., Pauwels, L., Plovie, E., Inzé, D., Gheysen, G., & Goossens, A. (2009). Expression of the Arabidopsis jasmonate signalling repressor JAZ1/TIFY10A is stimulated by auxin. *EMBO Reports*, *10*, 923–928.

Harms, K., Ramirez, I., & Penacortes, H. (1998). Inhibition of wound-induced accumulation of allene oxide synthase transcripts in flax leaves by aspirin and salicylic acid. *Plant Physiology*, *118*, 1057–1065.

Horváth, E., Szalai, G., & Janda, T. (2007). Induction of abiotic stress tolerance by salicylic acid signaling. *Journal of Plant Growth Regulation*, *26*, 290–300.

Hou, X., Lee, L. Y. C., Xia, K., Yan, Y., & Yu, H. (2010). DELLAs modulate jasmonate signaling via competitive binding to JAZs. *Developmental Cell*, *19*, 884–894.

Hu, Y., Jiang, Y., Han, X., Wang, H., Pan, J., & Yu, D. (2017). Jasmonate regulates leaf senescence and tolerance to cold stress: Crosstalk with other phytohormones. *Journal of Experimental Botany*, *68*, 1361–1369.

Ichimura, K., Mizoguchi, T., Yoshida, R., Yuasa, T., & Shinozaki, K. (2000). Various abiotic stresses rapidly activate Arabidopsis MAP kinases AtMPK4 and AtMPK6. *Plant Journal*, *24*, 655–665.

Jang, G., Yoon, Y., & Choi, Y. D. (2020). Crosstalk with jasmonic acid integrates multiple responses in plant development. *International Journal of Molecular Sciences*, *21*(1), 305.

Jiang, Y., Liang, G., Yang, S., & Yu, D. (2014). Arabidopsis WRKY57 functions as a node of convergence for jasmonic acid- and auxin-mediated signaling in jasmonic acid-induced leaf senescence. *Plant Cell*, *26*, 230–245.

Katsir, L., Chung, H. S., Koo, A. J., & Howe, G. A. (2008). Jasmonate signaling: A conserved mechanism of hormone sensing. *Current Opinion in Plant Biology*, *11*, 428–435.

Kazan, K. (2015). Diverse roles of jasmonates and ethylene in abiotic stress tolerance. *Trends in Plant Science*, *20*, 219–229.

Kazan, K., & Manners, J. M. (2012). JAZ repressors and the orchestration of phytohormone crosstalk. *Trends in Plant Science*, *17*, 22–31.

Khan, M. I. R., Syeed, S., Nazar, R., & Anjum, N. A. (2012). An insight into the role of salicylic acid and jasmonic acid in salt stress tolerance. In *Phytohormones and Abiotic Stress Tolerance in Plants* (pp. 277–300). Springer, Berlin, Heidelberg.

Khan, S., & Stone, J. M. (2007). Arabidopsis thaliana GH3.9 in auxin and jasmonate cross talk. *Plant Signaling & Behavior*, *2*, 483–485.

Kim, J. H., Chung, K. M., & Woo, H. R. (2011). Three positive regulators of leaf senescence in Arabidopsis, ORE1, ORE3, and ORE9: Play roles in crosstalk among multiple hormone-mediated senescence pathways. *Gene & Genomics*, *33*, 373–381.

Klay, I., Pirrello, J., Riahi, L., Bernadac, A., Cherif, A., Bouzayen, M., et al. (2014). Ethylene response factor Sl-ERF.B.3 is responsive to abiotic stresses and mediates salt and cold stress response regulation in tomato. *Scientific World Journal*, *2014*, 167681.

Ku, Y. S., Sintaha, M., Cheung, M. Y., & Lam, H. M. (2018). Plant hormone signaling crosstalks between biotic and abiotic stress responses. *International Journal of Molecular Sciences*, *19*(10), 3206.

Lackman, P., González-Guzmán, M., Tilleman, S., Carqueijeiro, I., Pérez, A. C., Moses, T., Seo, M., Kanno, Y., Häkkinen, S. T., Van Montagu, M. C., & Thevelein, J. M. (2011). Jasmonate signaling involves the abscisic acid receptor PYL4 to regulate metabolic reprogramming in Arabidopsis and tobacco. *Proceedings of the National Academy of Sciences*, *108*, 5891–5896.

Laudert, D., & Weiler, E. W. (1998). Allene oxide synthase: A major control point in Arabidopsis thaliana octadecanoid signalling. *Plant Journal*, *15*, 675–684.

Leon-Reyes, A., Du, Y., Koornneef, A., Proietti, S., Körbes, A. P., Memelink, J., Pieterse, C. M., & Ritsema, T. (2010). Ethylene signaling renders the jasmonate response of Arabidopsis insensitive to future suppression by salicylic acid. *Molecular Plant-Microbe Interactions*, *23*, 187–197.

Lorenzo, O., Piqueras, R., Sánchez-Serrano, J. J., & Solano, R. (2003). Ethylene response factor1 integrates signals from ethylene and jasmonate pathways in plant defense. *Plant Cell*, *15*, 165–178.

Lyons, R., Manners, J. M., & Kazan, K. (2013). Jasmonate biosynthesis and signaling in monocots: A comparative overview. *Plant Cell Reports*, *32*, 815–827.

McSteen, P., & Zhao, Y. (2008). Plant hormones and signaling: Common themes and new developments. *Developmental Cell*, *14*(4), 467–473.

Miura, K., & Tada, Y. (2014). Regulation of water, salinity, and cold stress responses by salicylic acid. *Frontiers in Plant Science*, *5*, 4.

Moons, A., De Keyser, A., & Van Montagu, M. (1997). A group 3 LEA cDNA of rice, responsive to abscisic acid, but not to jasmonic acid, shows variety-specific differences in salt stress response. *Gene*, *191*, 197–204.

Mur, L. A., Prats, E., Pierre, S., Hall, M. A., & Hebelstrup, K. H. (2013). Integrating nitric oxide into salicylic acid and jasmonic acid/ethylene plant defense pathways. *Frontiers in Plant Science*, *4*, 215.

Mur, L. A., Sivakumaran, A., Mandon, J., Cristescu, S. M., Harren, F. J., & Hebelstrup, K. H. (2012). Haemoglobin modulates salicylate and jasmonate/ethylene-mediated resistance mechanisms against pathogens. *Journal of Experimental Botany*, *63*, 4375–4387.

Naik, G. R., Mukherjee, I., & Reid, D. M. (2002). Influence of cytokinins on the methyl jasmonate-promoted senescence in *Helianthus annuus* cotyledons. *Plant Growth Regulation*, *38*, 61–68.

Ndamukong, I., Abdallat, A. A., Thurow, C., Fode, B., Zander, M., Weigel, R., & Gatz, C. (2007). SA-inducible Arabidopsis glutaredoxin interacts with TGA factors and suppresses JA-responsive PDF1.2 transcription. *Plant Journal*, *50*, 128–139.

Pal, M., Szalai, G., Kovacs, V., Gondor, O. K., & Janda, T. (2013). Salicylic acid-mediated abiotic stress tolerance. In S. Hayat, A. Ahmed, & M. N. Alyemeni (Eds.), *Plant Growth and Development* (pp. 183–247). Springer.

Palmieri, M. C., Sell, S., Huang, X., Scherf, M., Werner, T., Durner, J., & Lindermayr, C. (2008). Nitric oxide-responsive genes and promoters in *Arabidopsis thaliana*: A bioinformatics approach. *Journal of Experimental Botany*, *59*, 177–186.

Pauwels, L., Barbero, G. F., Geerinck, J., Tilleman, S., Grunewald, W., Pérez, A. C., Chico, J. M., Bossche, R. V., Sewell, J., Gil, E., & García-Casado, G. (2010). NINJA connects the co-repressor TOPLESS to jasmonate signaling. *Nature*, *464*, 788–791.

Per, T. S., Khan, M. I. R., Anjum, N. A., Masood, A., Hussain, S. J., & Khan, N. A. (2018). Jasmonates in plants under abiotic stresses: Crosstalk with other phytohormones matters. *Environmental and Experimental Botany*, *145*, 104–120.

Pieterse, C. M., Van der Does, D., Zamioudis, C., Leon-Reyes, A., & Van Wees, S. C. (2012). Hormonal modulation of plant immunity. *Annual Review of Cell and Developmental Biology*, *28*, 489–521.

Poltronieri, P., Taurino, M., Domenico, S. D., Bonsegna, S., & Santino, A. (2013). Activation of the jasmonate biosynthetic pathway in roots in drought stress. In N. Tuteja & S. S. Gill (Eds.), *Climate Change and Abiotic Stress Tolerance* (pp. 325–342). Wiley-VCH Verlag GmbH.

Proietti, S., Bertini, L., Timperio, A. M., Zolla, L., Caporale, C., & Caruso, C. (2013). Crosstalk between salicylic acid and jasmonate in *Arabidopsis* investigated by an integrated proteomic and transcriptomic approach. *Molecular Biosystems*, *9*, 1169–1187.

Qi, T., Huang, H., Wu, D., Yan, J., Qi, Y., Song, S., & Xie, D. (2014). Arabidopsis DELLA and JAZ proteins bind the WD-repeat/bHLH/MYB complex to modulate gibberellin and jasmonate signaling synergy. *Plant Cell*, *26*, 1118–1133.

Ren, C., Han, C., Peng, W., Huang, Y., Peng, Z., Xiong, X., Zhu, Q., Gao, B., & Xie, D. (2009). A leaky mutation in DWARF4 reveals an antagonistic role of brassinosteroid in the inhibition of root growth by jasmonate in *Arabidopsis*. *Plant Physiology*, *151*, 1412–1420.

Robert-Seilaniantz, A., Navarro, L., Bari, R., & Jones, J. D. (2007). Pathological hormone imbalances. *Current Opinion in Plant Biology*, *10*, 372–379.

Roychoudhury, A., & Aftab, T. (2021). Phytohormones, plant growth regulators and signaling molecules: Cross-talk and stress responses. *Plant Cell Reports*, *40*, 1301–1303.

Sano, H., Seo, S., Koizumi, N., Niki, T., Iwamura, H., & Ohashi, Y. (1996). Regulation of cytokinins by endogenous levels of jasmonic and salicylic acids in mechanically wounded tobacco plants. *Plant Cell Physiology*, *37*, 762–769.

Sharabi-Schwager, M., Lers, A., Samach, A., & Guy, C. L. (2010). Overexpression of the CBF2 transcriptional activator in *Arabidopsis* delays leaf senescence and extends plant longevity. *Journal of Experimental Botany*, *61*, 261–273.

Skalak, J., Nicolas, K. L., Vankova, R., & Hejatko, J. (2021). Signal integration in plant abiotic stress responses via multistep phosphorelay signaling. *Frontiers in Plant Science*, *12*, 196.

Song, S., Choi, Y., Moon, Y. H., Kim, S., Choi, Y. D., & Lee, J. S. (2000). Systemic induction of a *Phytolacca insularis* antiviral protein gene by mechanical wounding, jasmonic acid, and abscisic acid. *Plant Molecular Biology*, *43*, 439–450.

Spoel, S. H., Koornneef, A., Claessens, S. M., Korzelius, J. P., van Pelt, J. A., Mueller, M. J., Buchala, A. J., Métraux, J. P., Brown, R., Kazan, K., & van Loon, L. C. (2003). NPR1 modulates cross-talk between salicylate- and jasmonate-dependent defense pathways through a novel function in the cytosol. *Plant Cell*, *15*, 760–770.

Sreenivasulu, N., Harshavardhan, V. T., Govind, G., Seiler, C., & Kohli, A. (2012). Contrapuntal role of ABA: Does it mediate stress tolerance or plant growth retardation under long-term drought stress? *Gene*, *506*, 265–273.

Staswick, P. E., & Tiryaki, I. (2004). The oxylipin signal jasmonic acid is activated by an enzyme that conjugates it to isoleucine in *Arabidopsis*. *Plant Cell*, *16*, 2117–2127.

Staswick, P. E., Su, W., & Howell, S. H. (1992). Methyl jasmonate inhibition of root growth and induction of a leaf protein are decreased in an *Arabidopsis thaliana* mutant. *Proceedings of the National Academy of Sciences*, *89*(15), 6837–6840.

Stoynova-Bakalova, E., Petrov, P. I., Gigova, L., & Baskin, T. I. (2008). Differential effects of methyl jasmonate on growth and division of etiolated zucchini cotyledons. *Plant Biology*, *10*, 476–484.

Takahashi, F., Yoshida, R., Ichimura, K., Mizoguchi, T., Seo, S., Yonezawa, M., Maruyama, K., Yamaguchi-Shinozaki, K., & Shinozaki, K. (2007). The mitogen-activated protein kinase cascade MKK3–MPK6 is an important part of the jasmonate signal transduction pathway in *Arabidopsis*. *Plant Cell*, *19*, 805–818.

Tiryaki, I. (2007). The role of auxin-signaling gene axr1 in salt stress and jasmonic acid inducible gene expression in *Arabidopsis thaliana*. *Journal of Molecular Cell Biology*, *6*, 189–195.

Tiryaki, I., & Staswick, P. E. (2002). An *Arabidopsis* mutant defective in jasmonate response is allelic to the auxin-signaling mutant axr1. *Plant Physiology*, *130*, 887–894.

van der Does, D., Leon-Reyes, A., Koornneef, A., Van Verk, M. C., Rodenburg, N., Pauwels, L., Goossens, A., Körbes, A. P., Memelink, J., Ritsema, T., & Van Wees, S. C. (2013). Salicylic acid suppresses jasmonic acid signaling downstream of SCFCOI1-JAZ by targeting GCC promoter motifs via transcription factor ORA59. *Plant Cell*, *25*, 744–761.

Verma, V., Ravindran, P., & Kumar, P. P. (2016). Plant hormone-mediated regulation of stress responses. *BMC Plant Biology*, *16*, 86.

Wasternack, C., & Strnad, M. (2016). Jasmonate signaling in plant stress responses and development—Active and inactive compounds. *New Biotechnology, 33*, 604–613.

Werner, T., Nehnevajova, E., Kollmer, I., Novak, O., Strnad, M., Kramer, U., et al. (2010). Root specific reduction of cytokinin causes enhanced root growth, drought tolerance, and leaf mineral enrichment in *Arabidopsis* and tobacco. *Plant Cell, 22*, 3905–3920.

Woo, H. R., Kim, H. J., Nam, H. G., & Lim, P. O. (2013). Plant leaf senescence and death-regulation by multiple layers of control and implications for aging in general. *Journal of Cell Science, 126*, 4823–4833.

Xin, L., Wuliang, S., Shuqiu, Z., & Chenghou, L. (2005). Nitric oxide involved in signal transduction of jasmonic acid-induced stomatal closure of *Vicia faba* L. *Chinese Science Bulletin, 50*, 520–525.

Xing, Y., Jia, W., & Zhang, J. (2008). AtMKK1 mediates ABA-induced CAT1 expression and H_2O_2 production via AtMPK6-coupled signaling in *Arabidopsis*. *Plant Journal, 54*, 440–451.

Yang, J., Duan, G., Li, C., Liu, L., Han, G., Zhang, Y., & Wang, C. (2019). The crosstalks between jasmonic acid and other plant hormone signaling highlight the involvement of jasmonic acid as a core component in plant response to biotic and abiotic stresses. *Frontiers in Plant Science, 10*, 1349.

Zhang, H., Huang, Z., Xie, B., Chen, Q., Tian, X., Zhang, X., et al. (2004). The ethylene-, jasmonate-, abscisic acid- and NaCl-responsive tomato transcription factor JERF1 modulates expression of GCC box-containing genes and salt tolerance in tobacco. *Planta, 220*, 262–270.

Zhang, X., Zhu, Z., An, F., Hao, D., Li, P., Song, J., Yi, C., & Guo, H. (2014). Jasmonate activated MYC2 represses ETHYLENE INSENSITIVE3 activity to antagonize ethylene promoted apical hook formation in *Arabidopsis*. *Plant Cell, 26*, 1105–1117.

Zhu, Z. (2014). Molecular basis for jasmonate and ethylene signal interactions in *Arabidopsis*. *Journal of Experimental Botany, 65*, 5743–5748.

Zhu, Z., & Lee, B. (2015). Friends or foes: New insights in jasmonate and ethylene co-actions. *Plant Cell Physiology, 56*, 414–420.

CHAPTER 6

Role of Jasmonates in Plant Growth and Development

ARJUMAND JOHN,[1] KAISAR AHMAD BHAT,[2] AASHIQ HUSSAIN MIR,[1] SAJAD A. GANIE,[3] KHURSHEED HUSSAIN,[4] and ALI ASGHAR SHAH[2]

[1]*Department of Zoology, Lovely Professional University, Phagwara, Jalandhar, Punjab, India*

[2]*School of Biosciences and Biotechnology, Baba Ghulam Shah Badshah University, Rajouri, Jammu and Kashmir, India*

[3]*Division of Entomology, Sheri-e-Kashmir University of Agricultural Sciences, Shalimar, Jammu and Kashmir, India*

[4]*Department of Botany, University of Kashmir Hazratbal, Srinagar, Jammu and Kashmir, India*

ABSTRACT

Jasmonic acid (JA) naturally occurring growth regulator found in higher plants. Growth and developmental processes in plants are influenced by transcription factors activating JA (Jasmonic Acid) signaling. As this phytohormone (JA) controls growth, development, and defence by fusing endogenous and external signals. This lipid-derived plant stress hormone controlsa variety of biological and abiotic issues, such as disease and herbivore assault. Jasmonates contribute to root development, reproductive development, and leaf senescence in the context of plant development and growth. Jasmonates can inhibit, promote, or induce a variety of physiological processes in plants. According to the information below, jasmonates are crucial for the development of plants and their growth.

Biotechnological Applications of Jasmonates: A Group of Less Well-Known Plant Growth Regulators.
Tariq Ahmad Dar, Bilal Ahmad, and Khalid Rehman Hakeem (Eds.)

6.1 INTRODUCTION

As a member of the ubiquitous cyclopentanone family, jasmonic acid (JA) is believed to serve as a new growth and development regulator. Plants respond to both biotic and abiotic stress using jasmonates, which are ubiquitously occurring lipid-derived signaling agents. Physiological effects are likely to be exerted by them in all plants. In other words, they are secondary products of plants. Interest in plant compounds influencing several physiological systems has recently grown (Table 6.1). Jasmonates have several impacts due to their pervasiveness in complex and subordinate plants (Meyer et al., 1984). Phytohormones have antagonistic and synergistic impacts on the development and growth of plants, including fungi and algae (Miersch et al., 1987). These chemicals have the ability to indirectly alter plant organs and tissues in a way similar to how changes in their endogenous jasmonate levels occur. Another factor that led some scientists to consider jasmonates as potential hormonally active chemicals was their chemical resemblance to prostaglandins. It is still debatable whether calculating the total amount of endogenous phytohormones without accounting for the sensitivity (or "competence" state of the target cells) justifies inferring that these chemicals are in charge of regulating a range of physiological consequences. Although plant growth regulators are implicated in signal transduction cascades for plant growth and growth processes, the huge amount of data demonstrating their molecular and cellular actions supports the notion that these substances act both intrinsically and extrinsically to control plant growth and development processes (Figure 6.1).

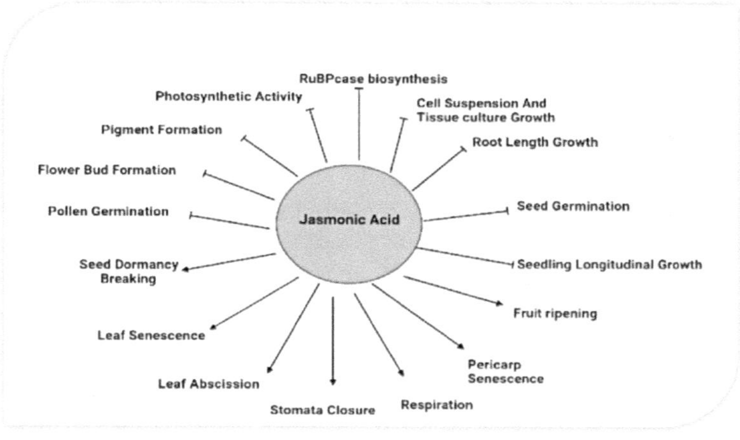

FIGURE 6.1 Role of jasmonates in plant growth and development.

6.2 BIOLOGICAL ACTIONS OF JASMONATES

Jasmonates, as one would anticipate from plant growth and development-regulating chemicals, have both inhibitory and indorsing (inducing) actions in plants or plant structures. Table 6.1 is a collection of the majority of the stated impacts. Jasmonates suppress growth-related processes while initiating, encouraging, or speeding up ontogenetic processes, such as organ ripening and senescence, at medium concentrations (10-4 to 10-6 M). Yamane et al. (1981) found that jasmonates inhibit seedling development and callus growth. Jasmonates impair embryogenesis and seed germination. Inhibit premature germination while inducing ESP (embryo-specific proteins) (Wilen et al., 1991). When seeds are desiccated, JA may act as a substitute or supplement to ABA in the developmental control of the seeds. Bryonia tendrils were discovered to coil in response to jasmonates

TABLE 6.1 Jasmonate Effect on Plant Growth and Developmental Processes

Physical Process	JA Effect	References
Longitudinal growth of seedlings	Inhibition	Yamane et al. (1980)
Root length growth	Inhibition	Yamane et al. (1981)
Tissue culture growth and cell suspension	Inhibition	Ueda and Kato (1982)
Seed Germination	Inhibition	Yamane et al. (1981)
Embryogenesis	Inhibition	Wilen et al. (1991)
Pollen germination	Inhibition	Yamane et al. (1981)
Flower bud formation	Inhibition	Barendse et al. (1985)
Pericarp senescence	Promotion	Lopez et al. (1987)
Fruit ripening	Promotion	Saniewski et al. (1987)
Seed Dormancy breaking	Promotion	Berestetsky et al. (1991)
Leaf senescence	Promotion	Ueda and Kato (1980)
Leaf abscission	Promotion	Curtis et al. (1984)
Tuber formation	Induction	Yoshihara et al. (1989)
Tendril coiling	Induction	Falkenstein et al. (1991)
Stomata closure	Promotion	Satler and Thimann (1981)
Microtubule disruption	Induction	Abe et al. (1990)
Pigment formation	Inhibition	Abeles et al. (1989)
Chlorophyll degradation	Promotion	Abeles et al. (1989)
Respiration	Promotion	Satler and Thimann (1981)
RuBPCase biosynthesis	Inhibition	Popova and Vaklinova (1988)
Photosynthetic activity	Inhibition	Popova et al. (1988)

and a-linolenic acid. Tuberonic acid, a 12-hydroxy-JA metabolite, and its O-beta glucoside have been shown to stimulate tuber development in potato stolon tips (Yoshihara et al., 1989). However, other jasmonate derivatives have limited activity compared to JA, CA, and their methyl esters (Koda et al., 1991). Tuber-inducing action depends on free acids' metabolism (hydroxylation and glucosylation) and their methyl esters, or on the overall delivery of the physiologically active structure.

6.2.1 JASMONIC ACID ACCUMULATION AND DISTRIBUTION

The concentration of JA varies between tissues and cells and is regulated by a wide range of environmental variables. Within 12 hours of imbibition, soybean seeds have JA levels of 1 g/g fresh weight while developing axes have levels of 2 g/g fresh weight. Compared to stems, leaves, and roots at more mature stages, seedlings exhibit greater concentrations of JA in the hypocotyl hook, a zone of cell division. To a minor range, epidermal cells display increased levels of jasmonate receptor expression in mature soybean leaves (Franceschi et al., 1983; Huang et al., 1991). When leaves are treated with jasmonate, palisade parenchyma cells express little jasmonate-responsive gene expression (Huang et al., 1991). It is likely that palisade cells have lower JA levels or compartmentalize JA differently or that these cells are more sensitive to JA. As palisade cells are filled with chloroplasts, their cytoplasm may contain fewer levels of JA than those needed to activate gene expression, since they are packed with chloroplasts. Increases in endogenous JA levels by overexpression of allene oxide synthase do not activate gene expression (Harms et al., 1995), as JA accumulates in chloroplasts. The concentration of JA in plant cells is likely to fluctuate transiently and dramatically in response to exogenous JA before reaching internal equilibrium. By contrast, higher rates of JA synthesis are achieved in chloroplasts when allene oxide synthase is overexpressed. Plants synthesize JA in chloroplast-containing cells that can sequester it as it is produced. JA receptors that modulate gene expression may also be located in other areas of the cell. The accumulation of JA during root growth could also be induced by mechanical impedance (Staswick et al., 1992). JA may also mediate growth changes induced by wind or touch. Mechanical perturbation induces upregulation of TCH genes in Arabidopsis thaliana (Braam et al., 1990). Calcium-binding proteins calmodulin and two others are encoded by TCH genes. These genes are more likely to be expressed in response to a rise in cytoplasmic calcium levels (Antosiewicz et al., 1995). Activation of phospholipase and lipoxygenase by calcium stimulates eicosanoid biosynthesis in animal cells. There are similar enzymes involved in the synthesis of JA in plant cells, suggesting that touch and turgor reduction may activate JA synthesis through a similar signal transduction pathway.

6.3 JASMONATE BIOSYNTHESIS AND STEREOCHEMISTRY

Jasmonic acid (JA) and precursors, which are created from cyclic fatty acids, control several plant defense and growth processes. Plants also conjugate JA with isoleucine to create (+)-7-iso-jasmonoyl-l-isoleucine (JA-Ile), which is a volatile derivative of JA known as methyl jasmonate (Me-JA). Together with other related compounds, these molecules are known as jasmonates (JAs). Plants and fungi use two types of jasmonic acid, (–)-JA and (+)-7-iso-JA (Figure 6.2). This cis-form of JA results in a molar equilibrium during trans-(–)-JA conversion by isomerization, as Vick and Zimmerman noted, 1984. Keto functions are present in the C-6 of derivatives of jasmonic acid. Cucurbic acid is reduced to a hydroxyl group in order to produce CAI and CAI derivatives (Fukui et al., 1977). There is, however, a greater restriction on its use than that of jasmonates.

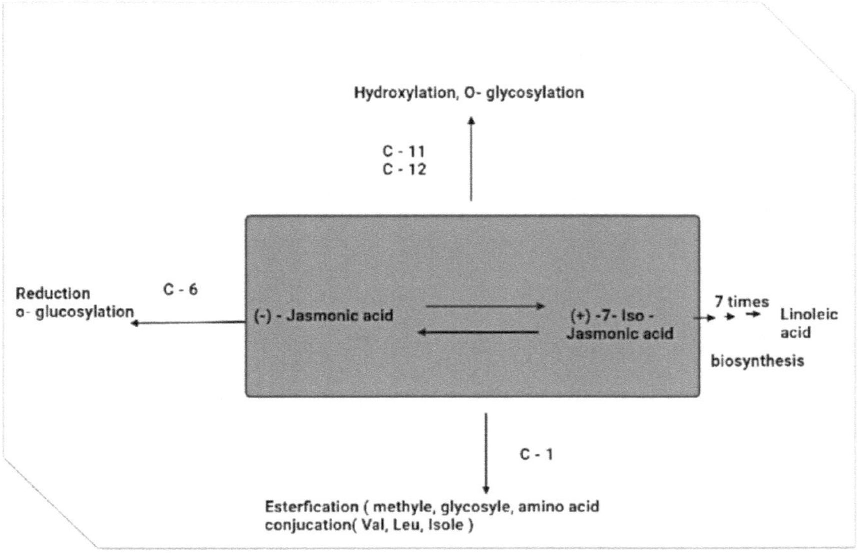

FIGURE 6.2 Jasmonate structures of two diastereomeric forms and their effects on metabolism.

There are three chemical moiety groups that make a jasmonoid substance biologically active: acetyl, n-pentenyl, and keto or hydroxy groups at C-6 (Yamane et al., 1980). Structures containing 9,11, O-dihydro-, 4,5-dihydro, and 3,7-dihydro-atoms are generally less active. In this case, the JA methyl ester appears to have a greater effect than the free acid, most likely because its absorption is rapid. Physicochemical and immunoassay techniques were used to identify JA and a number of its derivatives or conjugates (Sembdner et al., 1990). The structural variety of compounds has not yet been clarified whether most of them are biotransformation products or products of biosynthesis. Similar pathways may be involved in the metabolism of exogenously applied jasmonates. In Figure 6.2, there are three main routes for hydroxylation: C-12 or C-11, O-glucosylation, and conversion with some L-amino acids. Researchers found it interesting how (–) JA specificity affects amino acids. Aliphatic amino acids dominate the JA conjugates from Figure 6.2, as opposed to aromatic amino acids (phe, tyr, trp) found in the flowers and fruits of Vicia faba. Potential metabolic changes caused by physiologically active endogenous jasmonates. According to Meyer et al. (1991), amino acids in barley leaves produce JA conjugates (ile, leu, vale). Hydrophobicity appears to be essential for amino acids. Amino acid conjugates play a significant role in metabolism thanks to their (–)-JA moiety and L (S) form. Leaf tissues cannot split (R)-amino acids or (+)-JA conjugates and further metabolize them (R. Kramell). Jasmonates in particular, and conjugated phytohormones in general, have yet to be understood physiologically. The existence of jasmonates has been widely discussed even though they are phloem-mobile substances, but their validity has not yet been proven. Sembdner and Gross (1986) provide more information about the chemistry and occurrence of this compound.

6.4 DIFFERENT FUNCTIONS OF JASMONATES

6.4.1 HYPOCOTYL GROWTH INHIBITION

According to Chen et al. (2013), JA inhibits hypocotyl extension through COI1 in Arabidopsis. Far-red light greater than 640 nm enhances hypocotyl elongation in JA-deficient mutants and also suppresses hypocotyl growth in JA-deficient mutants (Robson et al., 2010). JA-mediated suppression of hypocotyl elongation under red light and in the dark does not affect hypocotyl length in COI1 mutant plants (Chen et al., 2013); their hypocotyls are longer under red light and in the dark. The myc2/jin1 mutant displays lengthened hypocotyls under far-red light and a low R/FR light ratio, but a shorter hypocotyl under blue light (Robson et al., 2010). In response to red and far-red light, MYC2 positively inhibits hypocotyl elongation, whereas blue light negatively inhibits it. Additionally, JA suppresses the development of ear shoots in Zea mays and rice (Coleoptiles and plant height are both inhibited by JA) (Biswas et al., 2003).

6.4.2 ROOT GROWTH INHIBITION

The co-receptors in *Arabidopsis thaliana* (Arabidopsis) recognize JA-Ile and JA imitate coronatine (Sheard et al., 2010). The majority of Arabidopsis JAZ members lack the amphiphilic repression domain linked to the ERF. EAR domains directly recruit TPLs and TRRs in some noncanonical JAZs (such as JAZ8 and JAZ13), whereas NINJA interacts with TPL and TPL-related proteins (TPRs) to recruit co-repressor TPLs and TPL-related proteins (TPLs) (Chini et al., 2007). The repression of primary root growth was lessened when the Jas domain in NINJA and JAZ proteins were deleted, altered, or changed. Combining mutations in JAZ7, JAZ8, JAZ10, and JAZ13 inhibited primary root growth in the absence of mutations in NINJA/TPL or overexpression of NINJA/TPL (Chini et al., 2007). MYC2, MYC3, MYC4, and MYC5 are a few Arabidopsis transcription factors that bind to and interact with JAZ proteins (Figure 6.3) (Fernandez-Calvo et al., 2011). Myc2s, Myc3s, and Myc4s, which are found near the tip of the primary roots, are the mediators via which JAs prevent the formation of primary roots (Fernandez-Calvo et al., 2011). MYC2, one of the components of the mediator complex, functions as a connection between TFs and RNA polymerase II (Cevik et al., 2012). PLETHORA (PLT)1 inhibits primary root growth, but PLT2 expression is suppressed by MYC2 according to Chen et al. (2011). The essential ethylene signaling proteins EIN3-LIKE1 (EIL1) and EIN3-INSENSITIVE (EIN)3 in Arabidopsis work together to promote the formation of root hair and the suppression of primary root growth by JA (Figure 6.3) (Zhu et al., 2011). The JA affects both lateral and adventitious roots. When JA is present, Arabidopsis does not generate adventitious roots (Gutierrez et al., 2012).

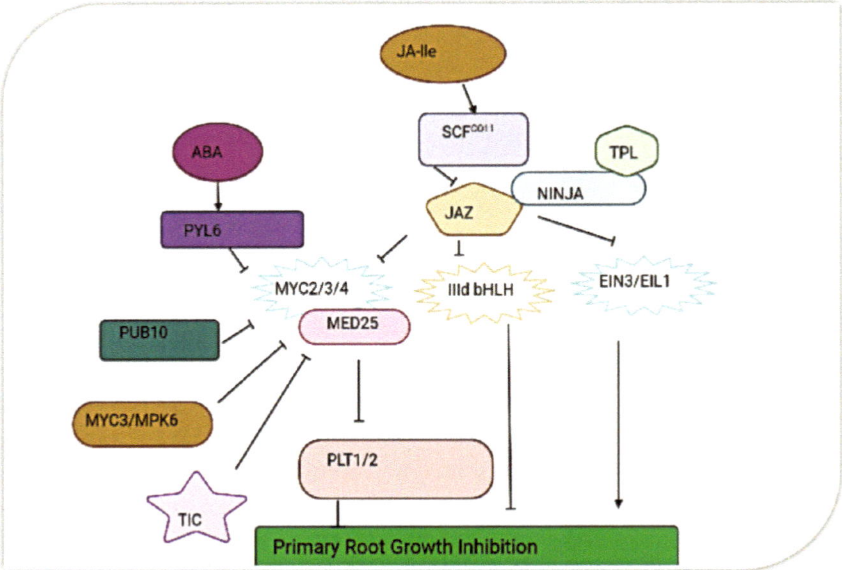

FIGURE 6.3 Regulating the inhibitory effect of JA on root growth.

6.4.3 LEAF SENESCENCE INDUCTION

Leaf senescence is induced by JA via a COI1-dependent mechanism (Qi et al., 2015). JA regulates leaf senescence induced by darkness. It is believed that leaf senescence occurs due to an increase in senescence genes (such as SAG29) and chlorophyll catabolism (such as Pheophorbide A Oxygenase) as well as a decrease in genes involved in photosynthesis (such as Chlorophyll A/B Binding Protein1) (Zhu et al., 2015). NAC019, NAC055, and NAC072 interact with CCGs to increase chlorophyll breakdown (Zhu et al., 2015). bHLH TFs in subgroup IIId delay leaf ageing by competing for binding to SAG29 (Qi et al., 2015). The mutants of *Myc2*, *Myc3*, and *Myc4* had a delayed senescence when tested with JA, but the mutants of BHDH17, BHDH13, BHDH3, and BHDH14 experienced accelerated senescence (Qi et al., 2015). According to Jiang et al. (2014), WRKY57 inhibits JA-induced leaf senescence and functions as a target of IAA-29 that studies the antagonistic relationship between JA and auxin (Figure 6.4). EIN3 is stabilized by ethanol through EIN2, and CCGs are activated by ORE1 to enhance leaf withering (Figure 6.4) (Zhang et al., 2016; Li et al., 2013).

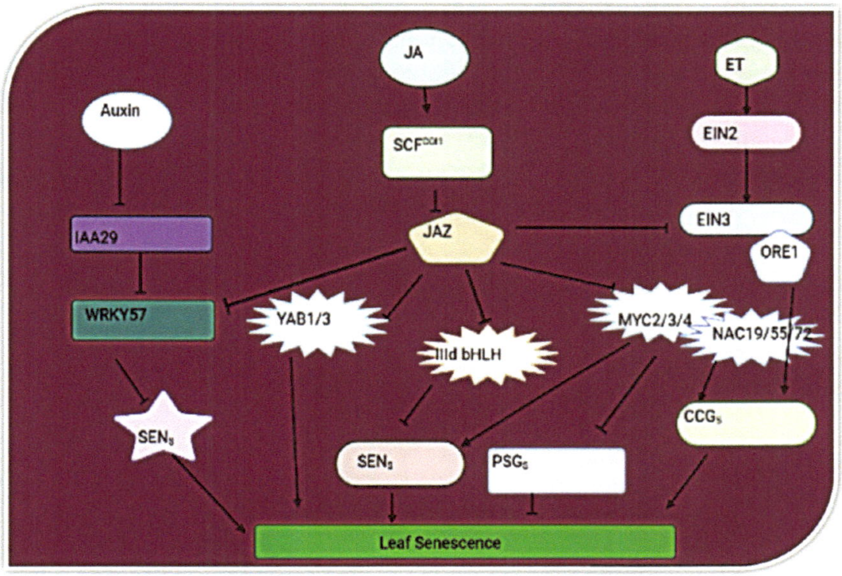

FIGURE 6.4 Involvement of JA in the regulation of leaf senescence.

6.5 DEVELOPMENT OF REPRODUCTIVE ORGANS IN PLANTS

6.5.1 CONTROL OF STAMEN DEVELOPMENT IN ARABIDOPSIS

Stamen plays a vital part in plant fertility. The JA protein is missing in multiple mutants of Arabidopsis (such as FAD3, AFAD7, and AFAD8), Lines overexpressing the catabolism gene CYP94B3; JA signaling mutant COI1; mutants in anther dehiscence1 (DAD1), lox3, lox4, aos, and opr3; as well as transgenic lines of JAZ13A and JAZ10.4 are male sterile as a consequence of stamen development being arrested at anthesis. Only acx1 and acx5 negatively impact pollen viability (Caldelari et al., 2011). Mutants lacking in JA signaling cannot be treated with JA, whereas those lacking in JA biosynthesis may. Anthers can dehisce, filaments can lengthen, and pollen can develop in a strain with restored COI1 expression (Jewell and Browse, 2016). JAZs directly target the MYB21, MYB24, and MYB57 R2R3-MYB TFs (Song et al., 2011). In JA-regulated stamen formation, several genes have distinct functions overlapping and overlapping at the same time (Figure 6.5; Cheng et al., 2009). They are completely sterile due to the short threads, late anther debridement, and untenable pollen grains. By upregulating MYB21, it was demonstrated that COI1 plants may regain stamen development (Song et al., 2011). According to Manda-okar and Browse (2009), MYB108 controls pollen maturation and anther dehiscence after MYB21. It has been shown that the IIIe bHLH factors physically interact with MYB21 and MYB24 to regulate the growth of stamens (Figure 6.5). (Qi et al., 2015). Stamen formation in Arabidopsis is dependent on the JA pathway, according to recent research (Song et al., 2013).

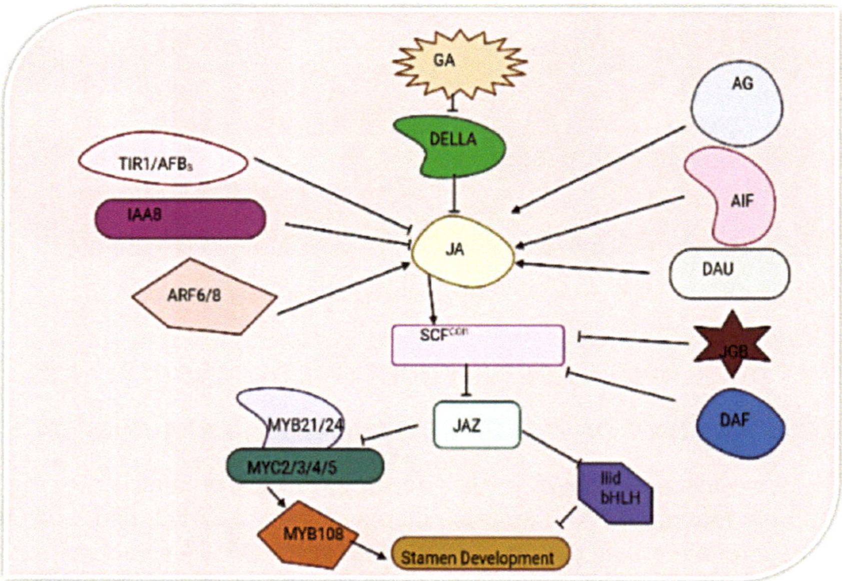

FIGURE 6.5 Cross-talk and JA signaling in stamen development.

6.5.2 DEVELOPMENT OF STAMEN IN ARABIDOPSIS

According to Song et al. (2013), mutations of JAZ13A and JAZ10.4 as well as overexpression lines of CYP94B3 (Jasmonoylisoleucine-12-hydroxylase) indicated impaired stamen development. Plants deficient in JA production are rescued by exogenous administration of JA, but mutants deficient in JA signaling is not (Jewell and Browse, 2016). Songs et al. (2011) report direct interactions between R2R3-MYB TFs MYB21, MYB24, and MYB57 and JAZs. Myb21 myb24 double mutants exhibit delayed anther dehiscence, non-viable pollen grains, and short filaments. Song et al. (2011) showed that overexpressing MYB21 proteins restored stamen formation in COI1-1 plants. MYB21 and MYB24 are physically connected to MYC2, MYC3, MYC4, and MYC5 which regulate stamen formation (Qi et al., 2015). Stamen maturation and productivity are enhanced in plants by overexpression of MYC5 and MYC3 (Qi et al., 2015).

6.5.3 EFFECTS ON SEED GERMINATION

Application of JA to seeds in Arabidopsis delays germination, although ABA-mediated inhibition by jar1 and COI1-16 has an impact (Ellis and Turner, 2002). Arabidopsis has an OPDA inhibitory mechanism that prevents seed germination that is COI1-independent (Dave et al., 2011). JA biosynthesis-related genes and JA biosynthesis are rapidly upregulated in latent embryos when wheat seeds (*Triticum aestivum*) are cold-stimulated to germinate. JA also inhibits ABA production to speed up germination (Xu et al., 2016).

6.5.4 ARABIDOPSIS APICAL HOOK FORMATION INHIBITION

Dark-grown seedlings have apical hooks that protect them upon emergence from the ground. (Song et al., 2014) contribute to JA's inability to form an apical hook via the COI1-JAZs-MYC2/3/4 cascade. A major function of JA is to activate MYC2 by interfering with and blocking EIN3/EIL1 transcription. Through down-regulation of the HOOKLESS1 gene, the apical hooks are prevented from curling (Zhang, 2014; Zhu, 2014). In fact, MYC2 promotes the production of the F-box protein 1 that binds to EIN3 and helps the degradation of the protein (Zhang et al., 2014). As ethylene stabilizes EIN3/EIL1, MYCs cannot perform their transcriptional function, thereby weakening the plant's defense against insect attack (Song et al., 2014).

6.5.5 INDUCTION OF SENESCENCE

A lack of endogenous jasmonates prevents senescent barley leaves from producing and accumulating JIPs; nevertheless, exogenous jasmonates cause connected senescent leaves to do so. Leaf abscission and physiological age are not significant variables for JA-induced JIP creation and accumulation because JIPs may also be produced in very young barley leaves. Senescence and VSP induction are not linked to soybean (Anderson et al., 1988). In cotton cotyledons, JIP production and yellowing symptoms don't happen (Reinbothe et al., 1991). Neither JA nor ethylene could induce leaf senescence (Abeles et al., 1989; Anderson, 1991). A study conducted by Saniewski et al. in 1987 found that JA-ME enhanced ethylene production in ripening tomatoes.

6.5.6 FLOWER DEVELOPMENT

Jasmonic acid (JA) biosynthesis and signaling play a crucial role in blooming. *Arabidopsis* mutants affected in JA biosynthesis, including *dad1*, *fad3-2 fad7-2 fad8*, *dde2-2*, *dde1*, *opr3*, and *aim1*, as well as mutants in JA signaling, such as *COI1*, exhibit clear JA-dependent phenotypes in flower development. In each of these mutants, anther development and dehiscence are delayed, or filament expansion is impaired (Turner et al. 2002; Delker et al. 2006). According to Xie et al. (1998) and Li et al. (2004), male sterility is induced in *Arabidopsis* by an impaired F-box protein *COI1*. The homolog in tomatoes, however, is crucial for female fertility. In double-mutant *arf6/arf8* filaments, JA levels are reduced, resulting in reduced filament elongation in JA-deficient mutants such as *dad1* and *opr3*. Two auxin response factors are impaired in these mutants, indicating that auxin signaling occurs upstream of JA (Nagpal et al. 2005). The *MYB21* and *MYB24* transcription factors are JA-inducible and influence the jasmonate response during stamen maturation, along with 13 other identified transcription factors. Hause et al. (2000) report specific functions for JA and other oxylipins in tomato flower organs based on the abundance of allene oxide cyclase (AOC) protein and the distinct oxylipin signatures. Meiersch et al. (2004) found that constitutive overexpression of *AOC* increases both 12-oxophytodienoic acid (OPDA) levels and the ratio of OPDA to JA. As our understanding of the oxylipin signature evolves, it will be interesting to see how it influences the development and abscission of flowers, the formation of embryos, and the development of seeds.

6.5.7 PETAL EXPANSION INHIBITION

Arabidopsis responds to jasmonate through the *COI1* pathway and suppresses petal growth. The petals of *COI1* mutant plants are larger at anthesis than those of the wild type. According to Brioudes et al. (2009) and Reeve et al. (2012), jasmonic acid (JA)-deficient mutants of *Arabidopsis* exhibited similar gene expression patterns. It has been demonstrated that *MYB21* (Myeloblastosis 21) and *MYB24* are essential for petal expansion, as JA represses their expression during the formation of sexual organs (Reeve et al. 2012). Plants from the *aos* and *COI1* families expressed more *MYB21* in their petals, leading to continuous petal expansion and the appearance of enlarged petals (Reeve et al., 2012). Furthermore, *opr3* plants express the basic helix-loop-helix (bHLH) transcription factor *BIGPETALp* at a higher level. In addition to regulating post-mitotic cell growth, this transcription factor also promotes petal expansion (Brioudes et al., 2009).

6.5.8 DEVELOPMENT OF FRUITS

Jasmonate may be involved in blooming, fruiting, and seeds since these processes occur in growing plant reproductive tissues, where relatively large quantities of this substance are present. The presence of jasmonate and similar volatile fatty acid derivatives contributes to insect attraction when pollen is distributed. The expression of genes encoding seeds, vegetative storage proteins, and fruit ripening are other factors that jasmonate can affect during the development of a flower, fruit, and seed. Jasmonate stimulates tomato and apple fruit ripening by activating EFE and producing ethylene (Czapski et al., 1992). Fruit ripening may be accelerated by an increase in jasmonate levels in developing fruit. By applying JA to tomato flesh, a smaller amount of lycopene was accumulated, while a greater amount of astaxanthin was accumulated (Saniewski et al., 1983). It is necessary to investigate the biochemical basis and physiological significance of this JA-mediated change.

6.6 RESISTANCE TO INSECTS AND DISEASES

A key role played by JA is in the resistance of plants to insects and diseases. This conclusion is supported by a number of evidence. The first finding is that JA accumulates in wounded plants (Creelman et al., 1992) as well as in plants treated with pathogen defense elicitors (Gundlach et al., 1992) that stimulate a plant's defense against pathogens. The second mechanism through which JA prevents plant damage is that it activates genes encoding protease inhibitors (Johnson et al., 1989). As well as thionine, osmotic, PDF, and *RIP60* genes, JA regulates the expression of genes encoding antifungal proteins. Proteins involved in the synthesis of barriers to infection such as PRP (Choi et al., 1994) are modulated by JA. JA induces phenolic oxidase (Deng et al., 1993), which is involved in plant defense, as well as genes involved in phytoalexin biosynthesis (Hildebrand et al., 1993). As well as providing JA, the oxylipin pathway also produces volatile aldehydes and alcohols that are essential to plant defence and wound healing. *Pseudomonas syringae* and *E coli* were completely inhibited by C6-aldehyde 2-hexenal (Avdiushko et al., 1995), while Aphid fecundity was reduced by C6-aldehydes and alcohols (Avdiushko et al., 1995). In the process of synthesis, hydroperoxy lyase acts on 13-hydroperoxylinolenic acid to produce these compounds. A rise in lipoxygenase activity and a rise in hydroperoxy lyase activity are induced by jasmonate, wounding, and elicitors (Grimes et al., 1992). By producing six-carbon compounds, plants enhance their ability to protect themselves.

6.7 DUAL ROLES OF JA IN DEVELOPMENT AND DEFENSE

Researchers made an effort to justify JA's dual function in plant growth and defense, in this section the following inquiries are posed to begin this discussion: Why are genes involved in insect defense, such as Pin2, and those that encode vegetative storage proteins controlled similarly? Why is JA inducible by injury or elicitation in older plant parts but high in young apical sinks, particularly reproductive structures? Pin2, the encoding gene for tomato protease inhibitors, and VspB, the gene for soybean vegetative storage protein, have strikingly similar expression regulation mechanisms. There is a high level of expression of both genes in the apical regions of vegetative plants as well as in flowers and reproductive tissues. In response to wounding and the application of JA, both genes are induced. It has been found that in illuminated plants, JA/wounding treatment significantly increases gene expression and that sugars and phosphate/auxin are synergistically inhibited by JA (DeWald et al., 1994). Plant cells often accumulate large amounts of the products of these two genes. Both genes are found in vacuoles. Since both of these genes are expressed similarly and their products are localized in similar locations in plants, it is suggested that they play nearly identical roles, except for the function of the encoded proteins. The fact that several proteins involved in plant defense also serve as vegetative storage proteins is not surprising for a variety of reasons. Vegetative storage proteins build up in completely hydrated plants with big vacuoles, in contrast to seed storage proteins, which must be stored at high density in practically desiccated seeds. As a result, there are fewer restrictions on VSP proteins than on seed storage proteins. Vegetative storage proteins and defense-related proteins must be accumulated in high quantities by plants without interfering with metabolism. In order to achieve this, vacuoles may be used to sequence these proteins. Furthermore, both types of proteins recover amino acids when they are no longer required for storage or defense. Plant defense proteins, particularly those stored in vacuoles and activated by consumption, are ideal storage proteins in the vegetative state because they are stored in vacuoles and activated by consumption. The majority of VSPs will likely be found to play a part in plant defense in the future since those VSPs that can also help with plant defense have added benefits for the plant. In vegetative apices, immature leaves, and reproductive structures, JA-responsive genes with elevated expressions, such as Pin2, may simply be a by-product of their function as vegetative storage proteins. The "optimal defense theory" put out by researchers in the field of chemical ecology is supported by the differential accumulation of defense-related chemicals in plant tissues of high value (Baldwin et al., 1994). According to this hypothesis, the defense should be given to plant sections that are very likely to be attacked and have a major impact on a plant's fitness. The "C/N hypothesis," a second category of chemical defense theory, places emphasis on the possibility that the allocation of carbon and nitrogen for defense might compete with their usage for development and growth (Baldwin et al., 1994). According to this idea, plants may produce nitrogen-rich defenses like proteins instead of carbon-rich defenses like tannins and phenols depending on the availability of nutrients. The C/N paradigm of chemical defense is congruent in a number of ways with the inducible or activated defense mechanism mediated by JA. Older leaves only allocate resources to defense when it is necessary because JA levels are higher and JA-inducible defense is triggered in response to injury or elicitors. Additionally, sugars as well as the presence of JA both control the expression of Pin2 (Johnson et al., 1990, Pena et al., 1992). Other genes implicated in plant defense, such as Chs, exhibit dual regulation by carbohydrates and JA. Furthermore, when plants are cultivated under restricted nitrogen circumstances, Vsp expression and perhaps genes like Pin2 are repressed (Staswick et al., 1991).

6.8 CONCLUSION

Numerous studies on JA functions have been done since the 1980s. It plays a crucial role in regulating plant growth, development, defense, and adaptability to external conditions. The results imply that JAs actively participate in growth-related developmental processes and that JAs control growth. In a range of plant species, JAs hinder seedling growth and promote leaf senescence. Plant JAs also control a number of reproductive and developmental pathways. Researchers can discover new functions for JA throughout plant growth, factors crucial to its activity, and the molecular basis for that action using enhanced genetic and biochemical methods. The stimulation of "defensive genes" and their expression in the plant body in response to or to mitigate stress, is believed to be mediated by a signal transduction pathway including jasmonates. Jasmonates may indeed be involved in a plant's survival strategy in this method by encouraging the sacrifice of a section of the plant. There is a new budding body of research examining the connection between JA and seed development, in addition to the roles of JA and other oxylipins that have been found thus far in flower development. New insights into cross-talk between signaling pathways following growth and development have been gained from transcription factors acting in JA signaling.

KEYWORDS

- ***Arabidopsis* mutants**
- **biochemical activity of jasmonates**
- **carbon compounds**
- **immunoassay techniques**
- **induction of senescence**
- **jasmonate action**
- **reproductive development**

REFERENCES

Abe, M., Shibaoka, H., Yamane, H., & Takahashi, N. (1990). Cell cycle-dependent disruption of microtubules by methyl jasmonate in tobacco BY-2 cells. *Protoplasma, 156*(1), 1–8.

Abeles, F. B., Hershberger, W. L., & Dunn, L. J. (1989). Hormonal regulation and intracellular localization of a 33-kD cationic peroxidase in excised cucumber cotyledons. *Plant Physiology, 89*(2), 664–668.

Anderson, J. M. (1988). Jasmonic acid-dependent increases in the level of specific polypeptides in soybean suspension cultures and seedlings. *Journal of Plant Growth Regulation, 7*(4), 203–211.

Anderson, J. M. (1991). Jasmonic acid-dependent increase in vegetative storage protein in soybean tissue cultures. *Journal of Plant Growth Regulation, 10*(1), 5–10.

Antosiewicz, D. M., Polisensky, D. H., & Braam, J. (1995). Cellular localization of the Ca^{2+} binding TCH3 protein of *Arabidopsis*. *The Plant Journal, 8*(5), 623–636.

Avdiushko, S., Croft, K. P., Brown, G. C., Jackson, D. M., Hamilton-Kemp, T. R., & Hildebrand, D. (1995). Effect of volatile methyl jasmonate on the oxylipin pathway in tobacco, cucumber, and *Arabidopsis*. *Plant Physiology, 109*(4), 1227–1230.

Baldwin, I. T. (1994). Chemical changes rapidly induced by folivory. *Insect-Plant Interactions, 5*, 1–23.

Barendse, G. W. M., Ceoes, A. F., Van den Ende, G., Bosveld, M., & Creemers, T. (1985). Role of hormones on flower bud formation in thin-layer explants of tobacco. *Biologia Plantarum, 27*(4), 408–412.

Berestetzky, V., Dathe, W., Daletskaya, T., Musatenko, L., & Sembdner, G. (1991). Jasmonic acid in seed dormancy of *Acer tataricum*. *Biochemie und Physiologie der Pflanzen, 187*(1), 13–19.

Biswas, K. K., Neumann, R., Haga, K., Yatoh, O., & Iino, M. (2003). Photomorphogenesis of rice seedlings: A mutant impaired in phytochrome-mediated inhibition of coleoptile growth. *Plant and Cell Physiology, 44*(3), 242–254.

Braam, J., & Davis, R. W. (1990). Rain-, wind-, and touch-induced expression of calmodulin and calmodulin-related genes in *Arabidopsis*. *Cell, 60*(3), 357–364.

Brioudes, F., Joly, C., Szécsi, J., Varaud, E., Leroux, J., Bellvert, F., & Bendahmane, M. (2009). Jasmonate controls late development stages of petal growth in *Arabidopsis thaliana*. *The Plant Journal, 60*(6), 1070–1080.

Caldelari, D., Wang, G., Farmer, E. E., & Dong, X. (2011). *Arabidopsis* lox3 lox4 double mutants are male sterile and defective in global proliferative arrest. *Plant Molecular Biology, 75*(1), 25–33.

Çevik, V., Kidd, B. N., Zhang, P., Hill, C., Kiddle, S., Denby, K. J., & Kazan, K. (2012). MEDIATOR25 acts as an integrative hub for the regulation of jasmonate-responsive gene expression in *Arabidopsis*. *Plant Physiology, 160*(1), 541–555.

Chehab, E. W., Yao, C., Henderson, Z., Kim, S., & Braam, J. (2012). *Arabidopsis* touch-induced morphogenesis is jasmonate mediated and protects against pests. *Current Biology, 22*(8), 701–706.

Chen, J., Sonobe, K., Ogawa, N., Masuda, S., Nagatani, A., Kobayashi, Y., & Ohta, H. (2013). Inhibition of *Arabidopsis* hypocotyl elongation by jasmonates is enhanced under

red light in a phytochrome B-dependent manner. *Journal of Plant Research, 126*(1), 161–168.
Chen, R., Jiang, H., Li, L., Zhai, Q., Qi, L., Zhou, W., & Li, C. (2012). The *Arabidopsis* mediator subunit MED25 differentially regulates jasmonate and abscisic acid signaling by interacting with the MYC2 and ABI5 transcription factors. *The Plant Cell, 24*(7), 2898–2916.
Cheng, H., Song, S., Xiao, L., Soo, H. M., Cheng, Z., Xie, D., & Peng, J. (2009). Gibberellin acts through jasmonate to control the expression of MYB21, MYB24, and MYB57 to promote stamen filament growth in *Arabidopsis*. *PLoS Genetics, 5*(3), e1000440.
Chini, A., Fonseca, S. G. D. C., Fernandez, G., Adie, B., Chico, J. M., Lorenzo, O., & Solano, R. (2007). The JAZ family of repressors is the missing link in jasmonate signaling. *Nature, 448*(7154), 666–671.
Choi, D., Bostock, R. M., Avdiushko, S., & Hildebrand, D. F. (1994). Lipid-derived signals that discriminate wound-and pathogen-responsive isoprenoid pathways in plants: Methyl jasmonate and the fungal elicitor arachidonic acid induce different 3-hydroxy-3-methylglutaryl-coenzyme A reductase genes and antimicrobial isoprenoids in *Solanum tuberosum* L. *Proceedings of the National Academy of Sciences, 91*(6), 2329–2333.
Creelman, R. A., Tierney, M. L., & Mullet, J. E. (1992). Jasmonic acid/methyl jasmonate accumulates in wounded soybean hypocotyls and modulates wound gene expression. *Proceedings of the National Academy of Sciences, 89*(11), 4938–4941.
Curtis, R. W. (1984). Abscission-inducing properties of methyl jasmonate, ABA, and ABA-methyl ester and their interactions with ethephon, AgNO3, and malformin. *Journal of Plant Growth Regulation, 3*(1), 157-–168.
Czapski, J., & Saniewski, M. (1992). Stimulation of ethylene production and ethylene-forming enzyme activity in fruits of the non-ripening nor and rin tomato mutants by methyl jasmonate. *Journal of Plant Physiology, 139*(3), 265–268.
Dave, A., Hernández, M. L., He, Z., Andriotis, V. M., Vaistij, F. E., Larson, T. R., & Graham, I. A. (2011). 12-Oxo-phytodienoic acid accumulation during seed development represses seed germination in *Arabidopsis*. *The Plant Cell, 23*(2), 583–599.
Delker, C., Stenzel, I., Hause, B., Miersch, O., Feussner, I., & Wasternack, C. (2006). Jasmonate biosynthesis in *Arabidopsis thaliana*—Enzymes, products, regulation. *Plant Biology, 8*(3), 297–306.
Deng, W., Hamilton-Kemp, T. R., Nielsen, M. T., Andersen, R. A., Collins, G. B., & Hildebrand, D. F. (1993). Effects of six-carbon aldehydes and alcohols on bacterial proliferation. *Journal of Agricultural and Food Chemistry, 41*(3), 506–510.
DeWald, D. B., Sadka, A., & Mullet, J. E. (1994). Sucrose modulation of soybean Vsp gene expression is inhibited by auxin. *Plant Physiology, 104*(2), 439–444.
Ellis, C., & Turner, J. G. (2002). A conditionally fertile *coi1* allele indicates cross-talk between plant hormone signaling pathways in *Arabidopsis thaliana* seeds and young seedlings. *Planta, 215*(4), 549–556.
Falkenstein, E., Groth, B., Mithöfer, A., & Weiler, E. W. (1991). Methyl jasmonate and α-linolenic acid are potent inducers of tendril coiling. *Planta, 185*, 316–322.
Fernández-Calvo, P., Chini, A., Fernández-Barbero, G., Chico, J. M., Gimenez-Ibanez, S., Geerinck, J., & Solano, R. (2011). The Arabidopsis bHLH transcription factors MYC3 and MYC4 are targets of JAZ repressors and act additively with MYC2 in the activation of jasmonate responses. *The Plant Cell, 23*(2), 701–715.

Franceschi, V. R., & Giaquinta, R. T. (1983). The paraveinal mesophyll of soybean leaves in relation to assimilate transfer and compartmentation: II. Structural, metabolic and compartmental changes during reproductive growth. *Planta, 157*, 422–431.

Fukui, H., Koshimizu, K., Yamazaki, Y., & Usuda, S. (1977). Structures of plant growth inhibitors in seeds of *Cucurbita pepo* L. *Agricultural and Biological Chemistry, 41*(1), 189–194.

Grimes, H. D., Koetje, D. S., & Franceschi, V. R. (1992). Expression, activity, and cellular accumulation of methyl jasmonate-responsive lipoxygenase in soybean seedlings. *Plant Physiology, 100*(1), 433–443.

Gundlach, H., Müller, M. J., Kutchan, T. M., & Zenk, M. H. (1992). Jasmonic acid is a signal transducer in elicitor-induced plant cell cultures. *Proceedings of the National Academy of Sciences, 89*(6), 2389–2393.

Gutierrez, L., Mongelard, G., Floková, K., Păcurar, D. I., Novák, O., Staswick, P., & Bellini, C. (2012). Auxin controls *Arabidopsis* adventitious root initiation by regulating jasmonic acid homeostasis. *The Plant Cell, 24*(6), 2515–2527.

Harms, K., Atzorn, R., Brash, A., Kuhn, H., Wasternack, C., Willmitzer, L., & Pena-Cortes, H. (1995). Expression of a flax allene oxide synthase cDNA leads to increased endogenous jasmonic acid (JA) levels in transgenic potato plants but not to a corresponding activation of JA-responding genes. *The Plant Cell, 7*(10), 1645–1654.

Hause, B., Stenzel, I., Miersch, O., Maucher, H., Kramell, R., Ziegler, J., & Wasternack, C. (2000). Tissue-specific oxylipin signature of tomato flowers: Allene oxide cyclase is highly expressed in distinct flower organs and vascular bundles. *The Plant Journal, 24*(1), 113–126.

Hildebrand, D. F., Brown, G. C., Jackson, D. M., & Hamilton-Kemp, T. R. (1993). Effects of some leaf-emitted volatile compounds on aphid population increase. *Journal of Chemical Ecology, 19*, 1875–1887.

Huang, J. F., Bantroch, D. J., Greenwood, J. S., & Staswick, P. E. (1991). Methyl jasmonate treatment eliminates cell-specific expression of vegetative storage protein genes in soybean leaves. *Plant Physiology, 97*(4), 1512–1520.

Jewell, J. B., & Browse, J. (2016). Epidermal jasmonate perception is sufficient for all aspects of jasmonate-mediated male fertility in *Arabidopsis*. *The Plant Journal, 85*(5), 634–647.

Jiang, Y., Liang, G., Yang, S., & Yu, D. (2014). Arabidopsis WRKY57 functions as a node of convergence for jasmonic acid—and auxin-mediated signalling in jasmonic acid–induced leaf senescence. *The Plant Cell, 26*(1), 230–245.

Johnson, R., & Ryan, C. A. (1990). Wound-inducible potato inhibitor II genes: Enhancement of expression by sucrose. *Plant Molecular Biology, 14*, 527–536.

Johnson, R., Narvaez, J., An, G., & Ryan, C. (1989). Expression of proteinase inhibitors I and II in transgenic tobacco plants: Effects on natural defense against *Manduca sexta* larvae. *Proceedings of the National Academy of Sciences, 86*(24), 9871–9875.

Kim, J., Chang, C., & Tucker, M. L. (2015). To grow old: Regulatory role of ethylene and jasmonic acid in senescence. *Frontiers in Plant Science, 6*, 20.

Koda, Y., Kikuta, Y., Tazaki, H., Tsujino, Y., Sakamura, S., & Yoshihara, T. (1991). Potato tuber-inducing activities of jasmonic acid and related compounds. *Phytochemistry, 30*(5), 1435–1438.

Li, L., Zhao, Y., McCaig, B. C., Wingerd, B. A., Wang, J., Whalon, M. E., & Howe, G. A. (2004). The tomato homolog of coronatine-insensitive1 is required for the maternal control of seed maturation, jasmonate-signaled defense responses, and glandular trichome development. *The Plant Cell, 16*(1), 126–143.

Li, Z., Peng, J., Wen, X., & Guo, H. (2013). Ethylene-insensitive3 is a senescence-associated gene that accelerates age-dependent leaf senescence by directly repressing miR164 transcription in *Arabidopsis*. *The Plant Cell, 25*(9), 3311–3328.

Lopez, R., Dathe, W., Brickner, C., Miersch, O., & Sembdner, G. (1987). Jasmonic acid in different parts of the developing soybean fruit. *Biochemical Physiology of Plants, 182*, 195–201.

Mandaokar, A., & Browse, J. (2009). MYB108 acts together with MYB24 to regulate jasmonate-mediated stamen maturation in *Arabidopsis*. *Plant Physiology, 149*(2), 851–862.

Mandaokar, A., Thines, B., Shin, B., Lange, B. M., Choi, G., Koo, Y. J., & Browse, J. (2006). Transcriptional regulators of stamen development in *Arabidopsis* identified by transcriptional profiling. *The Plant Journal, 46*(6), 984–1008.

Meyer, A., Miersch, O., Büttner, C., Dathe, W., & Sembdner, G. (1984). Occurrence of the plant growth regulator jasmonic acid in plants. *Journal of Plant Growth Regulation, 3*(1), 1–8.

Meyer, A., Schmidt, J., Gross, D., Jensen, E., Rudolph, A., Vorkefeld, S., & Sembdner, G. (1991). Amino acid conjugates as metabolites of the plant growth regulator dihydro jasmonic acid in barley (*Hordeum vulgare*). *Journal of Plant Growth Regulation, 10*(1), 17–25.

Miersch, O., Preiss, A., Sembdner, G., & Schreiber, K. (1987). (+)-7-iso-jasmonic acid and related compounds from *Botryodiplodia theobromine*. *Phytochemistry, 26*(4), 1037–1039.

Miersch, O., Weichert, H., Stenzel, I., Hause, B., Maucher, H., Feussner, I., & Wasternack, C. (2004). Constitutive overexpression of allene oxide cyclase in tomato (*Lycopersicon esculentum* cv. Lukullus) elevates levels of some jasmonates and octadecanoids in flower organs but not in leaves. *Phytochemistry, 65*(7), 847–856.

Nagpal, P., Ellis, C. M., Weber, H., Ploense, S. E., Barkawi, L. S., Guilfoyle, T. J., & Reed, J. W. (2005). Auxin response factors ARF6 and ARF8 promote jasmonic acid production and flower maturation. *The Plant Journal, 44*(2), 209–223.

Peña-Cortés, H., Liu, X., Serrano, J. S., Schmid, R., & Willmitzer, L. (1992). Factors affecting gene expression of patatin and proteinase-inhibitor-II gene families in detached potato leaves: Implications for their co-expression in developing tubers. *Planta, 186*, 495–502.

Popova, L. P., & Vaklinova, S. G. (1988). Effect of jasmonic acid on the synthesis of ribulose-1, 5-bisphosphate carboxylase-oxygenase in barley leaves. *Journal of Plant Physiology, 133*(2), 210–215.

Popova, L. P., Tsonev, T. D., & Vaklinova, S. G. (1988). Changes in some photosynthetic and photorespiratory properties in barley leaves after treatment with jasmonic acid. *Journal of Plant Physiology, 132*(3), 257–261.

Qi, T., Huang, H., Song, S., & Xie, D. (2015). Regulation of jasmonate-mediated stamen development and seed production by a bHLH-MYB complex in *Arabidopsis*. *The Plant Cell, 27*(6), 1620–1633.

Reeves, P. H., Ellis, C. M., Ploense, S. E., Wu, M. F., Yadav, V., Tholl, D., & Reed, J. W. (2012). A regulatory network for coordinated flower maturation. *PLoS Genetics, 8*(2), e1002506.

Reinbothe, S., Machmudova, A., Wasternack, C., Reinbothe, C., & Parthier, B. (1992). Jasmonate-induced proteins in cotton: Immunological relationship to the respective barley proteins and homology of transcripts to late embryogenesis abundant (Lea) mRNAs. *Journal of Plant Growth Regulation, 11*(1), 7–14.

Robson, F., Okamoto, H., Patrick, E., Harris, S. R., Wasternack, C., Brearley, C., & Turner, J. G. (2010). Jasmonate and phytochrome A signaling in *Arabidopsis* wound and shade responses are integrated through JAZ1 stability. *The Plant Cell, 22*(4), 1143–1160.

Saniewski, M., & Czapski, J. (1983). The effect of methyl jasmonate on lycopene and β-carotene accumulation in ripening red tomatoes. *Experientia, 39*, 1373–1374.

Saniewski, M., Nowacki, J., & Czapski, J. (1987). The effect of methyl jasmonate on ethylene production and ethylene-forming enzyme activity in tomatoes. *Journal of Plant Physiology, 129*(1, 2), 175–180.

Satler, S. O. (1981). Le jasmonate de méthyle: Nouveau et puissant promoteur de la sénescence des feuilles. *Plant Physiology, 66*(2), 246–249.

Sembdner, G., & Gross, D. (1986). Plant growth substances of plant and microbial origin. In *Plant Growth Substances 1985* (pp. 139–147).

Sembdner, G., Meyer, A., Miersch, O., & Bruckner, C. (1990). Metabolism of jasmonic acid. In *Plant Growth Substances 1988* (pp. 374–379).

Sheard, L. B., Tan, X., Mao, H., Withers, J., Ben-Nissan, G., Hinds, T. R., & Zheng, N. (2010). Jasmonate perception by inositol-phosphate-potentiated COI1–JAZ co-receptor. *Nature, 468*(7322), 400–405.

Song, S., Huang, H., Gao, H., Wang, J., Wu, D., Liu, X., & Xie, D. (2014). Interaction between MYC2 and ETHYLENE INSENSITIVE3 modulates antagonism between jasmonate and ethylene signaling in *Arabidopsis*. *The Plant Cell, 26*(1), 263–279.

Song, S., Qi, T., Huang, H., & Xie, D. (2013). Regulation of stamen development by coordinated actions of jasmonate, auxin, and gibberellin in *Arabidopsis*. *Molecular Plant, 6*(4), 1065–1073.

Song, S., Qi, T., Huang, H., Ren, Q., Wu, D., Chang, C., & Xie, D. (2011). The jasmonate-ZIM domain proteins interact with the R2R3-MYB transcription factors MYB21 and MYB24 to affect jasmonate-regulated stamen development in *Arabidopsis*. *The Plant Cell, 23*(3), 1000–1013.

Staswick, P. E., Huang, J. F., & Rhee, Y. (1991). Nitrogen and methyl jasmonate induction of soybean vegetative storage protein genes. *Plant Physiology, 96*(1), 130–136.

Staswick, P. E., Su, W., & Howell, S. H. (1992). Methyl jasmonate inhibition of root growth and induction of a leaf protein are decreased in an *Arabidopsis thaliana* mutant. *Proceedings of the National Academy of Sciences, 89*(15), 6837–6840.

Turner, J. G., Ellis, C., & Devoto, A. (2002). The jasmonate signal pathway. *The Plant Cell, 14*(suppl_1), S153–S164.

Ueda, J., & Kato, J. (1980). Isolation and identification of a senescence-promoting substance from wormwood (*Artemisia absinthium* L.). *Plant Physiology, 66*(2), 246–249.

Ueda, J., & Kato, J. (1982). Inhibition of cytokinin-induced plant growth by jasmonic acid and its methyl ester. *Physiologia Plantarum, 54*(3), 249–252.

Vick, B. A., & Zimmerman, D. C. (1984). Biosynthesis of jasmonic acid by several plant species. *Plant Physiology, 75*(2), 458–461.

Wilen, R. W., van Rooijen, G. J., Pearce, D. W., Pharis, R. P., Holbrook, L. A., & Moloney, M. M. (1991). Effects of jasmonic acid on embryo-specific processes in *Brassica* and *Linum* oilseeds. *Plant Physiology, 95*(2), 399–405.

Xie, D. X., Feys, B. F., James, S., Nieto-Rostro, M., & Turner, J. G. (1998). COI1: An *Arabidopsis* gene required for jasmonate-regulated defense and fertility. *Science, 280*(5366), 1091–1094.

Xu, Q., Truong, T. T., Barrero, J. M., Jacobsen, J. V., Hocart, C. H., & Gubler, F. (2016). A role for jasmonates in the release of dormancy by cold stratification in wheat. *Journal of Experimental Botany, 67*(11), 3497–3508.

Yamane, H., Sugawara, J., Suzuki, Y., Shimamura, E., & Takahashi, N. (1980). Syntheses of jasmonic acid-related compounds and their structure-activity relationships on the growth of rice seedlings. *Agricultural and Biological Chemistry, 44*(12), 2857–2864.

Yamane, H., Takagi, H., Abe, H., Yokota, T., & Takahashi, N. (1981). Identification of jasmonic acid in three species of higher plants and its biological activities. *Plant and Cell Physiology, 22*(4), 689–697.

Yoshihara, T., Omir, E. S. A., Koshino, H., Sakamura, S., Kkuta, Y., & Koda, Y. (1989). Structure of a tuber-inducing stimulus from potato leaves (*Solanum tuberosum* L.). *Agricultural and Biological Chemistry, 53*(10), 2835–2837.

Zhai, Q., Zhang, X., Wu, F., Feng, H., Deng, L., Xu, L., & Li, C. (2015). Transcriptional mechanism of jasmonate receptor COI1-mediated delay of flowering time in *Arabidopsis*. *The Plant Cell, 27*(10), 2814–2828.

Zhang, B., Wu, S., Zhang, Y. E., Xu, T., Guo, F., Tang, H., & Xue, Y. (2016). A high temperature-dependent mitochondrial lipase EXTRA GLUME1 promotes floral phenotypic robustness against temperature fluctuation in rice (*Oryza sativa* L.). *PLoS Genetics, 12*(7), e1006152.

Zhang, X., Zhu, Z., An, F., Hao, D., Li, P., Song, J., & Guo, H. (2014). Jasmonate-activated MYC2 represses ethylene insensitive3 activity to antagonize ethylene-promoted apical hook formation in *Arabidopsis*. *The Plant Cell, 26*(3), 1105–1117.

Zhang, Y. I., & Turner, J. G. (2008). Wound-induced endogenous jasmonates stunt plant growth by inhibiting mitosis. *PloS One, 3*(11), e3699.

Zhu, X., Chen, J., Xie, Z., Gao, J., Ren, G., Gao, S., & Kuai, B. (2015). Jasmonic acid promotes degreening via MYC2/3/4-and ANAC019/055/072-mediated regulation of major chlorophyll catabolic genes. *The Plant Journal, 84*(3), 597–610.

Zhu, Z. (2014). Molecular basis for jasmonate and ethylene signal interactions in *Arabidopsis*. *Journal of Experimental Botany, 65*(20), 5743–5748.

Zhu, Z., An, F., Feng, Y., Li, P., Xue, L., Jiang, Z., & Guo, H. (2011). The derepression of ethylene-stabilized transcription factors (EIN3/EIL1) mediates jasmonate and ethylene signaling synergy in *Arabidopsis*. *Proceedings of the National Academy of Sciences, 108*(30), 12539–12544.

CHAPTER 7

An overview of Jasmonate; biosynthesis and plant defense

TARIQ AHMAD DAR,[1] BILAL AHMAD,[2] UNJMAN RASHID,[3] BARKET ALI,[1] and K. R. HAKEEM[4]

[1]*PG Department of Botany, Sri Pratap College, Cluster University Srinagar, Jammu and Kashmir, India*

[2]*Higher Education Department Jammu and Kashmir, India*

[3]*Department of Botany, Jamia Hamdard University, New Delhi, India*

[4]*Department of Biological Sciences, Faculty of Science, King Abdulaziz University, Jeddah, Saudi Arabia*

ABSTRACT

Jasmonic acid including its derivatives like jasmonyl isoleucine and methyl jasmonate collectively referred to as jasmonates have emerged as novel plant growth regulators influencing plant physiological, biochemical and defense responses. Jasmonates are produced via the oxidation of poly-unsaturated fatty acids collectively called oxylipins. Jasmonates regulate interweaving plant responses against microbes, herbivores, mechanical damage, and other environmental stresses like abiotic stresses including UV-stress, osmotic stress, salt stress, cold stress, temperature stress, heavy metal stress, ozone stress, etc., the present review gives an overview about jasmonates, structure and biosynthesis, signaling, defense against biotic, abiotic, and heavy metal stress and their role in post-harvest management of fruits.

Biotechnological Applications of Jasmonates: A Group of Less Well-Known Plant Growth Regulators. Tariq Ahmad Dar, Bilal Ahmad, and Khalid Rehman Hakeem (Eds.)

7.1 INTRODUCTION

Plants unlike animals do not show locomotion and have to respond to external cues from their place. The environment around plants is not static but changes continuously. Therefore, plants need to adapt to the changing environment with some internal mechanisms. The most common one is the involvement of plant hormones to respond to the external environment either working alone or showing cross-talk with the signaling pathways of other plant hormones. Also, plant hormones play a vital role in alleviating the different stress conditions faced by plants like salt, drought, chilling, heat, heavy metal and other biotic stresses. Jasmonates the new class of plant hormones has been proven substantial to evade the harmful effects of stressful environment.

7.2 BIOSYNTHESIS AND SIGNALING

Jasmonates are the products of octadecanoid pathway synthesized in chloroplast and peroxisome by the oxidation of unsaturated fatty acids. The precursor of Jasmonates (JAs and MeJA) is localized in chloroplast membrane known as alpha linolinic acid. Phospholipases convert lipids into alpha linolenic acid, the addition of oxygen to alpha linolenic acid results in the formation of 13-HPOT by the action of enzyme 13-lipoxygenase (LOX). This 13-hydroperoxy-9, 11, 15-octadecatrienoic acid is converted by the enzyme allene oxide synthase to allene oxide. Allene oxide cyclase (AOC) converts allene oxide to 12- oxo phytodienoic acid (Figure 7.1) (Wasternack and Hause, 2013).

Jasmonate signaling in plants is induced by different elicitors like chitins, oligosaccharides, and different types of stresses like ozone (Rao et al., 2000) salt stress (Pedranzani et al., 2003), UV irradiation (Demkura et al., 2010) wounding, drought, osmotic (Kramell et al., 1995). Jasmonate signaling pathway is repressed by the Jasmonate-Zim Domain proteins which inactivates the transcriptional activators involved in it. Binding of jasmonoyl-isoleucine to coronatine insensitive 1 (COI1) the receptor present in SCF^{COI1} complex, causes degradation of Jasmonate-Zim Domain (JAZ) proteins through ubiquitination and subsequent degradation through 26S proteasome and activates the JA signaling (Figure 7.2). These findings have been confirmed from the studies on mutant coi 1 where from it was deduced that Skp/Cullin/F-box complex (SCF^{COI1} mediates jasmonate perception and signaling.

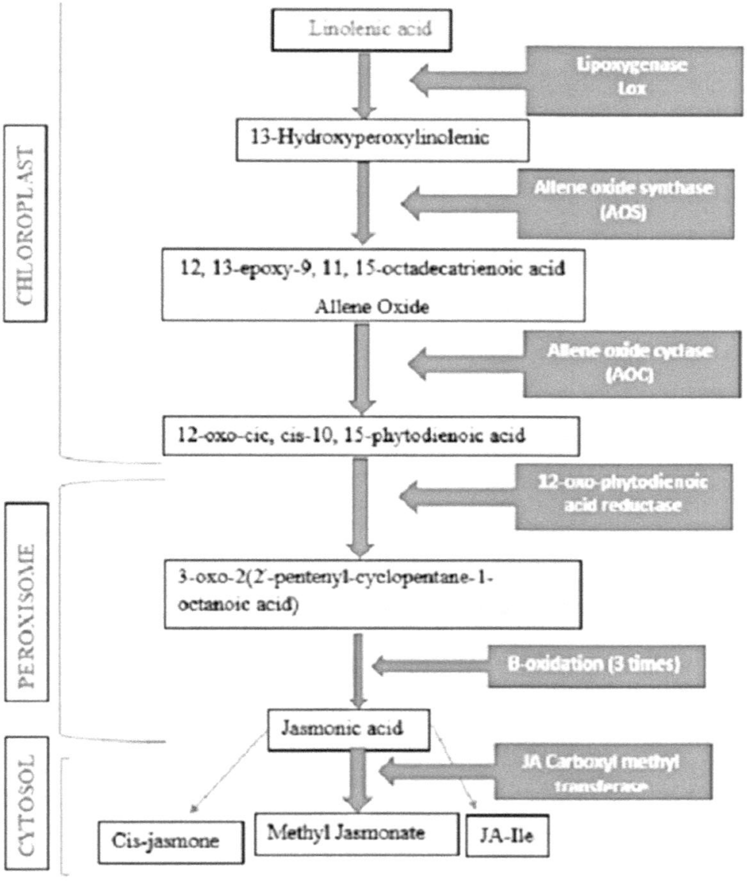

FIGURE 7.1 Biosynthetic Pathway of Jasmonates.

7.3 ABIOTIC STRESS RESPONSES

Among abiotic stresses faced by plants salinity stress is the major and most common influencing the growth, development, and yield of plants (Gharsallah et al., 2016). Salinity stress causes osmotic stress ionic imbalance and in plants which ultimately effects the biochemical processes, biomass, and morphology of plants (Rahneshan et al., 2018). Plants have adapted various strategies to mitigate the salinity stress and meanwhile Jasmonates have been found to alleviate the salinity stress in different plant species

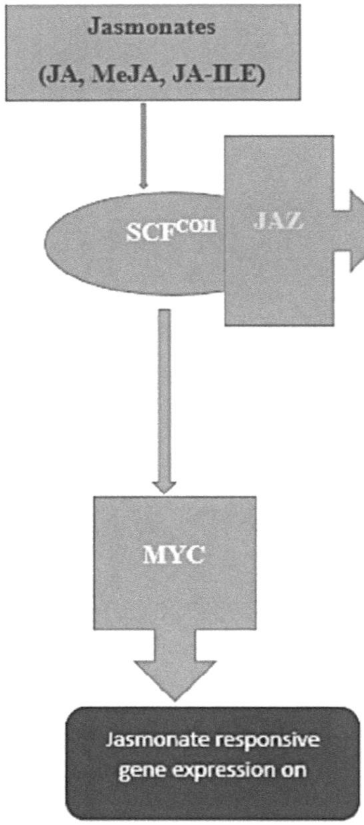

FIGURE 7.2 Jasmonate signaling pathway.

by different authors (Taheri et al., 2020; Riemann et al., 2015; Dar et al., 2015). Foliar application of MeJA comprehensively overcomes the symptoms in soybean seedlings as a result of salinity stress (Yoon et al., 2009). Priming with methyl jasmonate in *Pea* evades the salt stress and restored the fixation of CO_2 plant photosynthetic performance (Velitchkova and Fedina, 1998) and upregulated the MPK3 and G-beta subunit of G-proteins genes (Bhardwaj et al., 2011). Salt stress tolerance (Tsonev et al., 1998) in Barley seedlings improved with MeJA and showed less membrane damage (Bandurska and Stroinski, 2001). The endogenous Jasmonic acid content was observed to be enhanced in plants like arabidopsis (Ellouzi et al., 2013), tomato (Pedranzani et al., 2003), and potato (De Domenico et al., 2019) after salt treatment. The application of Jasmonates induces the synthesis of JIPs in response to different abiotic stresses (Sembdner and Parthier, 1993). Exogenous application of jasmonates enhanced the salt stress tolerance in *Robinia pseudoacacia* through enhancement in the activities of antioxidant enzymes superoxide dismutase and ascorbate peroxidase (Jiang et al., 2016). It was observed by Faghih et al. 2017 that Methyl jasmonate elevates the levels of Ascorbate peroxidase, peroxidase, and superoxide dismutase. MeJA treatment have been found to provide resistance against drought in different plants like groundnut (Todaka et al., 2015), rice (Fu et al., 2017), glycine max (Mohamed et al., 2017), and brassica oleraceae plants (Wu et al., 2012). The drought resistance was increased in *Vicia faba and Hordeum vulgare by* MeJA through regulating stomatal closure (Evans, 2003). In cauliflower drought resistance was increased by the application of methyl jasmonate through stimulation of enzymatic (Peroxidase, Superoxide dismutase, Catalase, Ascorbate peroxidase, and glutathione reductase), non-enzymatic (proline and soluble sugar) and molecules of antioxidative systems (Wu et al., 2012). Applying methyl jasmonate boosts the antioxidative defense system of plants. As reported by Abdelgawad et al. 2014 in maize plants jasmonates not only enhanced the level of osmolytes, carbohydrates, total soluble sugar, amino acid content, proline content, protein content but also the activities of antioxidative enzyme like catalase, peroxidase, and superoxide dismutase. MeJA effectively improves the tolerance of plants against drought by enhancing the antioxidative enzyme activity and osmoprotectants (Qiu et al., 2014). Methyl Jasmonate has been found to rouse the antioxidant compounds and enzymes (Figure 7.3) in different plants like *Prunus persica* (Yu et al., 2016), Eriobotrya (Rudell et al., 2007), Apple (Rudell et al., 2007), Brassica (Haroldsen et al., 2011), *Arabidopsis thaliana* (Sasaki-Sekimoto 2005), *Wolffia arrhiza* (Piotrowska et al., 2009), *Cajanus cajan* (Poonam et al., 2013), Kandelia (Chen et al., 2014), Prunus (Khana and Singha, 2007) Grapes (Flores et al., 2015.) Strawberry (Ayala-Zavala et al., 2005; Asghari and Hasanlooe, 2015), Blackberry (Wang et al., 2008; Huang et al., 2015), Pomegranates (Sayyari et al., 2011).

Higher levels of heavy metals in plants results in oxidative stress. JA have been found to alleviate the adverse effects of heavy metals in plants like in pigeon pea when seeds were treated with JA, the effect of Cu^{2+} on the growth of seedling was evaded and causes accumulation of carotenoids, Superoxide dismutase and Peroxidase (anti-oxidant enzymes) (Poonam et al., 2013) and proline. Cadmium stress in plants results in the production of reactive oxygen species (ROS) like hydroxyl radicals, hydrogen peroxide and superoxide radicals agents of oxidative damage in plant cells (Gallego et al., 2012). MeJA counters the cadmium stress in plants like *Soyabean, Arabidopsis thaliana, Capsicum frutescens* (Maksymiek and Krupa, 2002; Yan et al., 2009) *Kandelia obviate* (Chen et al., 2014). The genes of signaling pathways for alleviation of Cadmium stress are activated by MeJA and up-regulates the synthesis of Glutathione. Methyl Jasmonate is known to provide tolerance against Copper and Cadmium stress by the accumulation of phytochelatins in arabidopsis (Maksymiec et al., 2007).

It has been found that Methyl jasmonate application counteracts the chilling injury in the fruits of many plants like zucchini, magnifera, avocado, sweet pepper, Lycopersicon, and papaya (Gonzalez-Aguilar et al., 2000, 2003). Compared to the control plants when pineapple plants were treated with 10^{-3}, 10^{-4} and 10^{-5} M MeJA Chilling Injury index and electrolyte leakage were significantly reduced. In *Cucurbita pepo* MeJA reduces chilling injury through its regulation of ABA and polyamine levels (Wang and Buta, 1994) and delays the onset of chilling injury symptoms in. At chilling temperatures MeJA-supplied plants displayed an enhancement in abscissic acid synthesis and putrescine level with a simultaneous decrease in spermidine and spermine contents.

Heat shock proteins are synthesized in plants in response to high temperature that prevent denaturation of proteins. After application of methyl jasmonate at low-concentration in heat-stressed plants cell viability was restored as confirmed from electrolyte leakage studies. Accumulation of several jasmonate derivatives including 12-oxo-phytodienoic acid, Methyl jasmonate, Jasmonic acid and JA-Ile, and the expression of

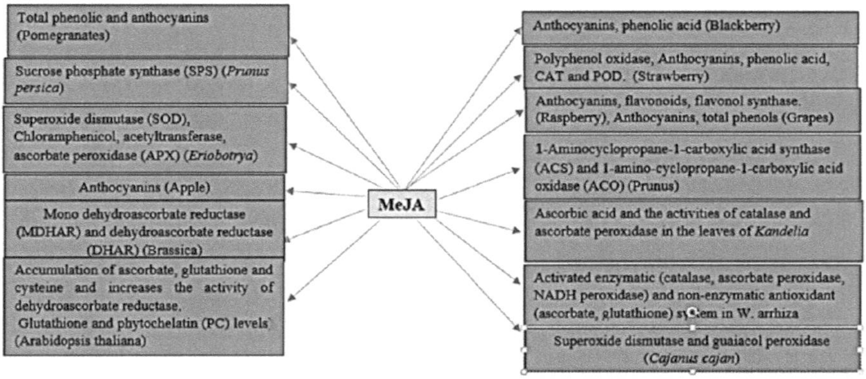

FIGURE 7.3 Role of methyl Jasmonate in activating the antioxidant system of plants.

jasmonate inducible gene PDF1.2 was found to be significantly enhanced upon heat stress exposure. Foliar application of MeJA on boron-stressed plants of Artemisia showed synthesis of antioxidant enzymes, reduces lipid peroxidation, and enhances artemisinin content.

7.4 JASMONATE MEDIATED BIOTIC STRESS TOLERANCE

Systemic acquired resistance in plants is induced by the exogenous application of MeJA and thereby loading them with a high-level protection against different biotic stresses including resistance to pathogens (Hayat et al., 2010). Application of jasmonates in field pots of tomato increase resistance of plants against pests (Thaler et al., 2001), protection against nematodes (Soriano et al., 2004) and against fungi in case of edible plants and vegetables. According to James 2005, Methyl Jasmonate and cis-jasmone have the herbivore repelling effect. Methyl Jasmonate and Jasmonic acid together enhanced the plant resistance against the green mold infection (Rohwer and Erwin, 2008). MeJA is known to effectively decrease the disease symptoms of grey mould rot in tomato fruit by controlling reactive oxygen species production (Zhu and Tian, 2012). In Peach fruits the MeJA treatment controls the post-harvest disease caused by *rhizopus stolonifer* by stimulating the defense enzymes like chitinase, β-1, 3-glucanase, PAL, polyphenol oxidase and peroxidase (Jin et al., 2009a). Methyl jasmonate rouse pathogen resistance through elevating the expression of MaWRKY1 gene (Tang et al., 2013). In wheat cultivar MeJA application significantly controls *Fusarium* crown and root rot by decreasing levels of hydrogen peroxide and lipid peroxidation, stimulating enzymatic activities and phenolic accumulation compounds (Motallebi et al., 2017). As reported by Kim et al., 2006 in *Ocimum,* methyl jasmonate fights pathogen infection with increase in the levels of terpenoids like eugenol and linalool and phenolic compounds like rosmarinic and caffeic acid. In spinach the survival of the fungus *Bradysia impatiens* was considerably reduced due to MeJA treatment (Schmelz et al., 2002). The population of *Myzus persicae* was significantly reduced by using of 5–10 mM of MeJA (Boughton et al., 2006). MeJA was found to be effective against nematodes in *Avena sativa* and *Spinacia oleracea* (Soriano et al., 2004). MeJA provides defense to the plants against insects in three major ways: (i) by activating antioxidant system of plants; (ii) Accumulation of defense related chemicals; and (iii) it can directly interact with insect predators.

7.5 ROLE OF JASMONATES IN PLANT GROWTH AND REPRODUCTIVE DEVELOPMENT

They are involved in several plant developmental processes which include stamen development, germination of seeds, ripening of fruits, root growth and senescence (Seltmann et al., 2010) storage organ formation, oxidative resistance (Per et al., 2018), elicitors of plant secondary metabolites (De Geyter et al., 2012), up regulation of antioxidant enzymes (Soares et al., 2010) systemic resistance (Pieterse et al., 2002, 2012). Jasmonates are known to promote the ethylene biosynthesis especially during fruit ripening in tomato and increases the activity of enzymes involved in the pathway viz ACC synthase and oxidase. The application of JA and octadecanoid pathway intermediates induces the tendril coiling response. They are also known to produce tubers in some crops like potato.

Treatments with jasmonates before as well as after harvesting may increase the synthesis of various secondary metabolites like phenolics, antioxidants, anthocyanins, flavonoids and enhancing the quality and post-harvest life of fruits. MeJA extend the shelf life of fruits and inhibits various post-harvest fungal diseases. MeJA treatments enhance the antioxidant activity in pomegranates (Sayyari et al., 2011) and blackberries (Wang et al., 2008).

The role of Jasmonates in plant defense responses against different stresses and promotion of growth, yield and reproductive development promises to pave the way in protection of crop plants and increase the crop productivity which ultimately leads to food security.

7.6 CONCLUSION

Different biotic and abiotic stresses affect both crop quality and quantity which cause losses regionally and globally. Therefore, plants need to identify and establish critical mechanisms to tolerate these harmful stresses and sustain their growth, development, and persistence. Jasmonates are the recently discovered plant hormones which participate in plant defense against abiotic and biotic stresses and modulate growth by the regulation of genes involved in stress responses. Tremendous amount of information has confirmed the role of methyl jasmonate in counteracting stress conditions under changing environments and activating the secondary metabolites and antioxidants including ascorbate/glutathione and glyoxalase. Also, jasmonates influence the various physiological processes in plants like seed germination, growth of primary roots and root hairs, fruit development, pollen maturation, senescence, and regeneration.

KEYWORDS

- **abiotic stress**
- **antioxidant enzymes**
- **biotic stress**
- **jasmonates**
- **oxylipins**
- **reproductive development**
- **signaling**

REFERENCES

Abdelgawad, Z. A., Khalafaallah, A. A., & Abdallah, M. M. (2014). Impact of methyl jasmonate on antioxidant activity and some biochemical aspects of maize plant grown under water stress condition. *Agricultural Sciences, 5*(12), 1077–1084.

Achuo, E. A., Audenaert, K., Meziane, H., & Höfte, M. (2004). The salicylic acid-dependent defense pathway is effective against different pathogens in tomato and tobacco. *Plant Pathology, 53*(1), 65–72.

Asghari, M., & Hasanlooe, A. R. (2016). Methyl jasmonate effectively enhanced some defense enzymes activity and total antioxidant content in harvested "Sabrosa" strawberry fruit. *Food Science & Nutrition, 4*(3), 377–383.

Ayala-Zavala, J. F., Wang, S. Y., Wang, C. Y., & González-Aguilar, G. A. (2005). Methyl jasmonate in conjunction with ethanol treatment increases antioxidant capacity, volatile compounds and postharvest life of strawberry fruit. *European Food Research and Technology, 221*(6), 731–738.

Bandurska, H., & Stroiński, A. (2001). Investigation of jasmonic acid participation in accumulation of ABA and proline and its role in response to water deficit in barley genotypes. *Acta Physiologiae Plantarum, 23*(3 Suppl.), 69–79.

Boughton, A. J., Hoover, K., & Felton, G. W. (2006). Impact of chemical elicitor applications on greenhouse tomato plants and population growth of the green peach aphid, *Myzus persicae*. *Entomologia Experimentalis et Applicata, 120*(3), 175–188.

Chen, J., Yan, Z., & Li, X. (2014). Effect of methyl jasmonate on cadmium uptake and antioxidative capacity in *Kandelia obovata* seedlings under cadmium stress. *Ecotoxicology and Environmental Safety, 104*, 349–356.

Dar, T. A., Uddin, M., Khan, M. M. A., Hakeem, K. R., & Jaleel, H. (2015). Jasmonates counter plant stress: A review. *Environmental and Experimental Botany, 115*, 49–57.

De Domenico, S., Taurino, M., Gallo, A., Poltronieri, P., Pastor, V., Flors, V., & Santino, A. (2019). Oxylipin dynamics in *Medicago truncatula* in response to salt and wounding stresses. *Physiologia Plantarum, 165*(2), 198–208.

De Geyter, N., Gholami, A., Goormachtig, S., & Goossens, A. (2012). Transcriptional machineries in jasmonate-elicited plant secondary metabolism. *Trends in Plant Science, 17*(6), 349–359.

Demkura, P. V., Abdala, G., Baldwin, I. T., & Ballaré, C. L. (2010). Jasmonate-dependent and -independent pathways mediate specific effects of solar ultraviolet B radiation on leaf phenolics and antiherbivore defense. *Plant Physiology, 152*(2), 1084–1095.

Doares, S. H., Syrovets, T., Weiler, E. W., & Ryan, C. A. (1995). Oligogalacturonides and chitosan activate plant defensive genes through the octadecanoid pathway. *Proceedings of the National Academy of Sciences, 92*(10), 4095–4098.

Ellouzi, H., Hamed, K. B., Cela, J., Müller, M., Abdelly, C., & Munné-Bosch, S. (2013). Increased sensitivity to salt stress in tocopherol-deficient *Arabidopsis* mutants growing in a hydroponic system. *Plant Signaling & Behavior, 8*(2), e23136.

Evans, N. H. (2003). Modulation of guard cell plasma membrane potassium currents by methyl jasmonate. *Plant Physiology, 131*(1), 8–11.

Faghih, S., Ghobadi, C., & Zarei, A. (2017). Response of strawberry plant cv. 'Camarosa' to salicylic acid and methyl jasmonate application under salt stress condition. *Journal of Plant Growth Regulation, 36*(3), 651–659.

Flores, G., Blanch, G. P., & del Castillo, M. L. R. (2015). Postharvest treatment with (−) and (+)-methyl jasmonate stimulates anthocyanin accumulation in grapes. *LWT-Food Science and Technology, 62*(1), 807–812.

Fu, J., Wu, H., Ma, S., Xiang, D., Liu, R., & Xiong, L. (2017). OsJAZ1 attenuates drought resistance by regulating JA and ABA signaling in rice. *Frontiers in Plant Science, 8*, 2108.

Gallego, S. M., Peña, L. B., Barcia, R. A., Azpilicueta, C. E., Iannone, M. F., Rosales, E. P., Zawoznik, M. S., Groppa, M. D., & Benavides, M. P. (2012). Unraveling cadmium toxicity and tolerance in plants: Insight into regulatory mechanisms. *Environmental and Experimental Botany, 83*, 33–46.

Gharsallah, C., Fakhfakh, H., Grubb, D., & Gorsane, F. (2016). Effect of salt stress on ion concentration, proline content, antioxidant enzyme activities and gene expression in tomato cultivars. *AoB Plants, 8*.

Gonzalez-Aguilar, G. A., Buta, J. G., & Wang, C. Y. (2003). Methyl jasmonate and modified atmosphere packaging (MAP) reduce decay and maintain postharvest quality of papaya 'Sunrise.' *Postharvest Biology and Technology, 28*(3), 361–370.
González-Aguilar, G. A., Fortiz, J., Cruz, R., Baez, R., & Wang, C. Y. (2000). Methyl jasmonate reduces chilling injury and maintains postharvest quality of mango fruit. *Journal of Agricultural and Food Chemistry, 48*(2), 515–519.
Haroldsen, V. M., Chi-Ham, C. L., Kulkarni, S., Lorence, A., & Bennett, A. B. (2011). Constitutively expressed DHAR and MDHAR influence fruit, but not foliar ascorbate levels in tomato. *Plant Physiology and Biochemistry, 49*(10), 1244–1249.
Hayat, Q., Hayat, S., Irfan, M., & Ahmad, A. (2010). Effect of exogenous salicylic acid under changing environment: A review. *Environmental and Experimental Botany, 68*(1), 14–25.
Horton, R. F. (1991). Methyl jasmonate and transpiration in barley. *Plant Physiology, 96*(4), 1376–1378.
Huang, X., Li, J., Shang, H., & Meng, X. (2015). Effect of methyl jasmonate on the anthocyanin content and antioxidant activity of blueberries during cold storage. *Journal of the Science of Food and Agriculture, 95*(2), 337–343.
James, D. G. (2005). Further field evaluation of synthetic herbivore-induced plant volatiles as attractants for beneficial insects. *Journal of Chemical Ecology, 31*(3), 481–495.
Jiang, M., Xu, F., Peng, M., Huang, F., & Meng, F. (2016). Methyl jasmonate regulated diploid and tetraploid black locust (*Robinia pseudoacacia* L.) tolerance to salt stress. *Acta Physiologiae Plantarum, 38*(4), 1–13.
Jin, P., Zheng, Y., Tang, S., Rui, H., & Wang, C. Y. (2009). Enhancing disease resistance in peach fruit with methyl jasmonate. *Journal of the Science of Food and Agriculture, 89*(5), 802–808.
Khan, A. S., & Singh, Z. (2007). Methyl jasmonate promotes fruit ripening and improves fruit quality in Japanese plum. *The Journal of Horticultural Science and Biotechnology, 82*(5), 695–706.
Kim, H. J., Chen, F., Wang, X., & Rajapakse, N. C. (2006). Effect of methyl jasmonate on secondary metabolites of sweet basil (*Ocimum basilicum* L.). *Journal of Agricultural and Food Chemistry, 54*(6), 2327–2332.
Kim, J. (2007). Perception, transduction, and networks in cold signaling. *Journal of Plant Biology, 50*(2), 139–147.
Kramell, R., Atzorn, R., Schneider, G., Miersch, O., Brückner, C., Schmidt, J., Sembdner, G., & Parthier, B. (1995). Occurrence and identification of jasmonic acid and its amino acid conjugates induced by osmotic stress in barley leaf tissue. *Journal of Plant Growth Regulation, 14*(1), 29–36.
Maksymiec, W., & Krupa, Z. (2002). Jasmonic acid and heavy metals in *Arabidopsis* plants—A similar physiological response to both stressors? *Journal of Plant Physiology, 159*(5), 509–515.
Maksymiec, W., Wojcik, M., & Krupa, Z. (2007). Variation in oxidative stress and photochemical activity in *Arabidopsis thaliana* leaves subjected to cadmium and excess copper in the presence or absence of jasmonate and ascorbate. *Chemosphere, 66*(3), 421–427.

Meir, S., Droby, S., Davidson, H., Alsevia, S., Cohen, L., Horev, B., & Philosoph-Hadas, S. (1998). Suppression of Botrytis rot in cut rose flowers by postharvest application of methyl jasmonate. *Postharvest Biology and Technology, 13*(3), 235–243.

Mohamed, H. I., & Latif, H. H. (2017). Improvement of drought tolerance of soybean plants by using methyl jasmonate. *Physiology and Molecular Biology of Plants, 23*(3), 545–556.

Motallebi, P., Niknam, V., Ebrahimzadeh, H., Hashemi, M., & Enferadi, S. T. (2017). Exogenous methyl jasmonate treatment induces defense response against *Fusarium culmorum* in wheat seedlings. *Journal of Plant Growth Regulation, 36*(1), 71–82.

Pedranzani, H., Racagni, G., Alemano, S., Miersch, O., Ramírez, I., Peña-Cortés, H., Taleisnik, E., Machado-Domenech, E., & Abdala, G. (2003). Salt tolerant tomato plants show increased levels of jasmonic acid. *Plant Growth Regulation, 41*(2), 149–158.

Pieterse, C. M., Schaller, A., Mauch-Mani, B., & Conrath, U. (2006). Signaling in plant resistance responses: Divergence and cross-talk of defense pathways. In *Multigenic and Induced Systemic Resistance in Plants* (pp. 166–196). Springer, Boston, MA.

Piotrowska, A., Bajguz, A., Godlewska-Żyłkiewicz, B., Czerpak, R., & Kamińska, M. (2009). Jasmonic acid as modulator of lead toxicity in the aquatic plant *Wolffia arrhiza* (Lemnaceae). *Environmental and Experimental Botany, 66*(3), 507–513.

Poonam, S., Kaur, H., & Geetika, S. (2013). Effect of jasmonic acid on photosynthetic pigments and stress markers in *Cajanus cajan* (L.) Millsp. seedlings under copper stress. *Environmental and Experimental Botany, 88*, 36–42.

Qiu, Z., Guo, J., Zhu, A., Zhang, L., & Zhang, M. (2014). Exogenous jasmonic acid can enhance tolerance of wheat seedlings to salt stress. *Ecotoxicology and Environmental Safety, 104*, 202–208.

Rahneshan, Z., Nasibi, F., & Moghadam, A. A. (2018). Effects of salinity stress on some growth, physiological, biochemical parameters and nutrients in two pistachio (*Pistacia vera* L.) rootstocks. *Journal of Plant Interactions, 13*(1), 73–82.

Rao, M. V., Lee, H. I., Creelman, R. A., Mullet, J. E., & Davis, K. R. (2000). Jasmonic acid signaling modulates ozone-induced hypersensitive cell death. *The Plant Cell, 12*(9), 1633–1646.

Riemann, M., Dhakarey, R., Hazman, M., Miro, B., Kohli, A., & Nick, P. (2015). Exploring jasmonates in the hormonal network of drought and salinity responses. *Frontiers in Plant Science, 6*, 1077.

Rohwer, C. L., & Erwin, J. E. (2008). Horticultural applications of jasmonates. *The Journal of Horticultural Science and Biotechnology, 83*(3), 283–304.

Rudell, D. R., Mattheis, J. P., Fan, X., & Fellman, J. K. (2002). Methyl jasmonate enhances anthocyanin accumulation and modifies production of phenolics and pigments in *Fuji* apples. *Journal of the American Society for Horticultural Science, 127*(3), 435–441.

Sasaki-Sekimoto, Y., Taki, N., Obayashi, T., Aono, M., Matsumoto, F., Sakurai, N., Suzuki, H., Hirai, M. Y., Noji, M., Saito, K., & Masuda, T. (2005). Coordinated activation of metabolic pathways for antioxidants and defense compounds by jasmonates and their roles in stress tolerance in *Arabidopsis*. *The Plant Journal, 44*(4), 653–668.

Sayyari, M., Babalar, M., Kalantari, S., Martínez-Romero, D., Guillén, F., Serrano, M., & Valero, D. (2011). Vapor treatments with methyl salicylate or methyl jasmonate alleviated chilling injury and enhanced antioxidant potential during postharvest storage of pomegranates. *Food Chemistry, 124*(3), 964–970.

Schmelz, E. A., Grebenok, R. J., Ohnmeiss, T. E., & Bowers, W. S. (2002). Interactions between *Spinacia oleracea* and *Bradysia impatiens*: A role for phytoecdysteroids. *Archives of Insect Biochemistry and Physiology, 51*(4), 204–221.

Seltmann, M. A., Stingl, N. E., Lautenschlaeger, J. K., Krischke, M., Mueller, M. J., & Berger, S. (2010). Differential impact of lipoxygenase 2 and jasmonates on natural and stress-induced senescence in *Arabidopsis*. *Plant Physiology, 152*(4), 1940–1950.

Sembdner, G. A. P. B., & Parthier, B. (1993). The biochemistry and the physiological and molecular actions of jasmonates. *Annual Review of Plant Biology, 44*(1), 569–589.

Soares, A. M. D. S., Souza, T. F. D., Jacinto, T., & Machado, O. L. T. (2010). Effect of methyl jasmonate on antioxidative enzyme activities and on the contents of ROS and H_2O_2 in *Ricinus communis* leaves. *Brazilian Journal of Plant Physiology, 22*, 151–158.

Soriano, I. R., Asenstorfer, R. E., Schmidt, O., & Riley, I. T. (2004). Inducible flavone in oats (*Avena sativa*) is a novel defense against plant-parasitic nematodes. *Phytopathology, 94*(11), 1207–1214.

Taheri, Z., Vatankhah, E., & Jafarian, V. (2020). Methyl jasmonate improves physiological and biochemical responses of *Anchusa italica* under salinity stress. *South African Journal of Botany, 130*, 375–382.

Tang, Y., Kuang, J. F., Wang, F. Y., Chen, L., Hong, K. Q., Xiao, Y. Y., Xie, H., Lu, W. J., & Chen, J. Y. (2013). Molecular characterization of PR and WRKY genes during SA- and MeJA-induced resistance against *Colletotrichum musae* in banana fruit. *Postharvest Biology and Technology, 79*, 62–68.

Thaler, J. S., Stout, M. J., Karban, R., & Duffey, S. S. (2001). Jasmonate-mediated induced plant resistance affects a community of herbivores. *Ecological Entomology, 26*(3), 312–324.

Todaka, D., Shinozaki, K., & Yamaguchi-Shinozaki, K. (2015). Recent advances in the dissection of drought-stress regulatory networks and strategies for development of drought-tolerant transgenic rice plants. *Frontiers in Plant Science, 6*, 84. https://doi.org/10.3389/fpls.2015.00084.

Tsonev, T. D., Lazova, G. N., Stoinova, Z. G., & Popova, L. P. (1998). A possible role for jasmonic acid in adaptation of barley seedlings to salinity stress. *Journal of Plant Growth Regulation, 17*(3), 153–159.

Velitchkova, M., & Fedina, I. (1998). Response of photosynthesis of *Pisum sativum* to salt stress as affected by methyl jasmonate. *Photosynthetica, 35*(1), 89–97.

Wang, C. Y., & Buta, J. G. (1994). Methyl jasmonate reduces chilling injury in *Cucurbita pepo* through its regulation of abscisic acid and polyamine levels. *Environmental and Experimental Botany, 34*(4), 427–432.

Wang, D., Weaver, N. D., Kesarwani, M., & Dong, X. (2005). Induction of protein secretory pathway is required for systemic acquired resistance. *Science, 308*(5724), 1036–1040.

Wang, S. Y., Bowman, L., & Ding, M. (2008). Methyl jasmonate enhances antioxidant activity and flavonoid content in blackberries (*Rubus sp.*) and promotes antiproliferation of human cancer cells. *Food Chemistry, 107*(3), 1261–1269.

Wasternack, C., & Hause, B. (2013). Jasmonates—Biosynthesis and role in stress responses and developmental processes. *Annals of Botany, 111*, 1021–1058.

Wasternack, C., & Hause, B. (2013). Jasmonates: Biosynthesis, perception, signal transduction and action in plant stress response, growth and development. An update to the 2007 review in *Annals of Botany, 111*(6), 1021–1058.

Wasternack, C., & Kombrink, E. (2010). Jasmonates: Structural requirements for lipid-derived signals active in plant stress responses and development. *ACS Chemical Biology, 5*(1), 63–77.
Wasternack, C., & Parthier, B. (1997). Jasmonate-signaled plant gene expression. *Trends in Plant Science, 2*(8), 302–307.
Wu, H., Wu, X., Li, Z., Duan, L., & Zhang, M. (2012). Physiological evaluation of drought stress tolerance and recovery in cauliflower (*Brassica oleracea* L.) seedlings treated with methyl jasmonate and coronatine. *Journal of Plant Growth Regulation, 31*(1), 113–123.
Yan, Z., Chen, J., & Li, X. (2013). Methyl jasmonate as modulator of Cd toxicity in *Capsicum frutescens* var. fasciculatum seedlings. *Ecotoxicology and Environmental Safety, 98*, 203–209.
Yoon, J. Y., Hamayun, M., Lee, S. K., & Lee, I. J. (2009). Methyl jasmonate alleviated salinity stress in soybean. *Journal of Crop Science and Biotechnology, 12*(2), 63–68.
Yu, J., Zhang, Y., Di, C., Zhang, Q., Zhang, K., Wang, C., You, Q., Yan, H., Dai, S. Y., Yuan, J. S., & Xu, W. (2016). JAZ7 negatively regulates dark-induced leaf senescence in *Arabidopsis*. *Journal of Experimental Botany, 67*(3), 751–762.
Zhu, Z., & Tian, S. (2012). Resistant responses of tomato fruit treated with exogenous methyl jasmonate to *Botrytis cinerea* infection. *Scientia Horticulturae, 142*, 38–43.

CHAPTER 8

Unravelling the Protective Role of Jasmonates During Heavy Metal Stress

SANTANU SAMANTA[1] and ARYADEEP ROYCHOUDHURY[2]

[1]*Post-Graduate Department of Biotechnology, St. Xavier's College (Autonomous), Mother Teresa Sarani, Kolkata, West Bengal, India*

[2]*Discipline of Life Sciences, School of Sciences, Indira Gandhi National Open University, Maidan Garhi, New Delhi, India*

ABSTRACT

Plants are constantly challenged by a myriad of abiotic stresses, which severely affect their growth and metabolism. Environmental pollution by heavy metals/metalloids poses a serious impact on crop yield losses worldwide. Heavy metal toxicity incites various toxicity symptoms like inhibition in biomass, germination potential, photosynthesis, alterations in physiological growth parameters, accumulation of excess reactive oxygen species, biochemical and metabolic attributes, culminating ultimately oxidative stress conditions. It is necessary to develop appropriate strategies in order to reduce such potentially toxic metals/metalloids in plant-derived foods. Jasmonates are lesser-known, lipid-derived plant growth regulators (PGRs) which provide protection during heavy metal/metalloid stresses in different plants. A detailed biosynthetic pathway of jasmonic acid and further downstream physiological regulations are discussed in this chapter, along with its complex signal transduction pathway in plants. The exogenous amendment of jasmonic acid and its methyl derivative antagonizes the adverse

Biotechnological Applications of Jasmonates: A Group of Less Well-Known Plant Growth Regulators.
Tariq Ahmad Dar, Bilal Ahmad, and Khalid Rehman Hakeem (Eds.)

effects of heavy metals/metalloids by regulating physiological and biochemical attributes in stressed plants. Here, we briefly describe the reported studies showing the crucial role of jasmonates to tackle heavy metal-provoked alterations at cellular level and their in-depth mechanisms, suggesting that metal(loid) stress tolerance might be improved by modulating jasmonate-mediated response. Additionally, the chapter also enlightens on cross-talk mechanisms orchestrated by jasmonates with other major phytohormones for enhancing tolerance under metal-toxicity conditions.

8.1 INTRODUCTION

Heavy metals are natural constituents of our planet and are associated with considerable disastrous effects even in meagre quantities for environmental biota. Extensive industrialization and various anthropogenic activities such as extensive application of fertilizers, industrial waste, smelting, and sewage disposal cause leaching of heavy metals into groundwater, which finally accumulate in the soil surface (Ghori et al., 2019). Generally heavy metals are non-biodegradable, possess relatively higher density and are toxic even at low concentrations. Build-up of heavy metals in soil is of great concern for crop cultivation due to entry into the food chain and subsequent accumulation in human body (Samanta and Roychoudhury, 2021a). Agricultural crop plants growing in heavy metal-infested soil exhibit altered photosynthesis and metabolism, growth retardation, and bioaccumulation of excess metals in the different plant organs. Their concentration beyond the threshold level is toxic to plants primarily due to uncontrolled generation of reactive oxygen species (ROS) which cause several toxic manifestations (Manara, 2012). Therefore, appropriate methods to mitigate the damages and crop loss imposed by heavy metal toxicity should be given top priority in the current scenario. Numerous conventional technologies are employed to cope with metal toxicity, but there are several major drawbacks to utilize these methods. An encouraging and eco-friendly approach in this regard is exogenous/foliar application or seed priming of plant growth regulators (PGRs) or phytohormones that are widely employed to combat heavy metal(loid) stress by regulating plant resilience and boost up stress tolerance.

Plant hormones often referred to as phytohormones act as chemical messengers that aid plants to promote stress endurance under varying environment (Roychoudhury et al., 2009, 2013). Since its discovery, methyl jasmonate (MeJA) and its free acid (JA) – collectively referred to as jasmonates, have gained attention of plant physiologists because of their potential role in plant defense (Cheong and Choi, 2003; Yu et al., 2018). They are a new group (class) of lesser-known PGRs due to their vast physiological roles and abundance in plants (Wang et al., 2020). In recent decades, exploring the functions of jasmonates in mitigating toxic metal stress has received intense research attention from many researchers in the field of plant-environment interactions. Its role as an anti-stressor against biotic stresses, like insect, pathogen, and plant herbivory has been well characterized (Avdiushko et al., 1997; Vijayan et al., 1998; Zavala and Baldwin, 2006). However, their involvement in enhancing abiotic stress tolerance, especially under metal/metalloid stress has been supported only recently either by exogenous supplementation or by utilizing seed priming/pre-sowing techniques. In this chapter, we summarize recent advances in research on the role of jasmonates in plant resistance mechanisms under heavy metal/metalloid stress. Furthermore, this chapter also suggested elaborated signaling cross-talk among other major phytohormones, which facilitates plants to integrate various abiotic stress responses.

8.2 SCHEME OF BIOSYNTHETIC PATHWAY OF JASMONIC ACID

Jasmonates are lipid-derived signaling molecules, act as important regulators in biotic and abiotic stresses as well as variety of physiological functions in plants. Jasmonates belong to cyclopentanone compounds and have two recognized forms: jasmonic acid (JA) and its volatile methyl ester [methyl jasmonates (MeJA)] widely distributed in the plant kingdom (Emamverdian et al., 2020). The biological roles of jasmonates are more or less similar to some of the well-known mediators in animals, most notably prostaglandins. In most plants, jasmonates are biosynthesized via octadecanoic pathway from the 18-carbon unsaturated fatty acid α-linoleic acid (Figure 8.1), while 20-carbon eicosanoid arachidonic acid acts as a precursor molecule for prostaglandins (Browse, 2009; Wasternack and Hause, 2013).

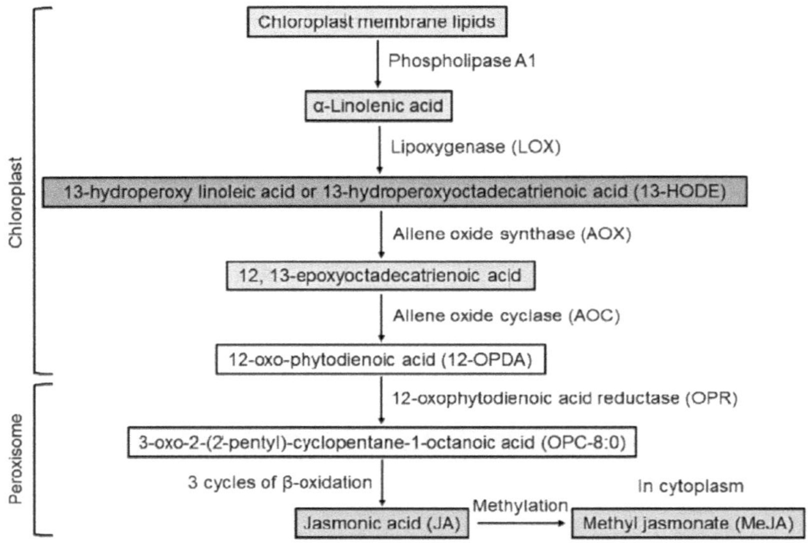

FIGURE 8.1 An outline of jasmonic acid biosynthetic pathway in plants.

Some plants also use 16-carbon fatty acids for the generation of jasmonates (Farmer et al., 1998). Linoleic acid is oxidized by lipoxygenase (LOX) to 13-hydroperoxylinoleic acid or 13-hydroperoxyoctadecatrienoic acid (13-HODE) in the plastid. Plants treated with LOX inhibitors or transgenic plants with down-regulated LOX activity have shown reduced production of jasmonates, which indicated that LOX plays an indispensable role in jasmonate biosynthesis (Mueller, 1997; León and Sánchez-Serrano, 1999). 13-HODE acts as a substrate for allene oxide synthase (AOS) and allene oxide cyclase (AOC) to form 12-oxo-phytodienoic acid (12-oxo PDA or 12-OPDA). For further conversion, 12-OPDA is exported from chloroplasts and entered into the peroxisomes through an ABC transporter (JAT1). In peroxisome, 12-OPDA is converted to 12-oxophytoenoic acid (OPC) by the action of OPDA reductase (OPR). Finally, carbon is removed from the carboxyl side chain via three rounds of β-oxidation steps for the generation of final product JA, which might be further metabolized.

8.3 METABOLITES DERIVED FROM JASMONATE

After biosynthesis, JA is exported from peroxisome to cytoplasm and subject to a variety of derivatives. JA can be converted to the biologically active form, 7-*iso*-JA-isoleucine (Ile), which is the amino acid-conjugated form of JA, catalyzed by the activity of Jasmonate Resistant 1 (JAR1). JAR1 is capable to conjugate amino acids to carboxyl groups of JA, like Ile, leucine, phenylalanine, tyrosine, tryptophan, and valine *in-vitro*, which is required for phytohormone activation (Fonseca et al., 2009; Kombrink, 2012). JA-Ile is the most widely available amino acid-conjugated (bioactive) natural form found in plants (Hause et al., 2000; Thines et al., 2007). MeJA is generated by accepting methyl group from S-adenosyl-1-methionine (SAM) to JA, by the action of enzyme JA-methyltransferase (Wasternack and Strnad, 2016). The methyl-esterified product (MeJA) is volatile in nature in contrast to JA. Exogenously supplemented MeJA is de-esterified to JA, which plays a key role in signal transduction under stress environments (Ghorbel et al., 2021). *Cis*-jasmonate, a decarboxylated derivative of jasmonate, is produced via oxidative decarboxylation reaction, which has been reported to be inactive in most of the bioassays in several plant species.

8.4 JASMONATE-MEDIATED RESPONSES IN PLANTS

Jasmonates are a group of less-known plant growth regulators involved in various aspects of plant life, such as root growth, gravitropism, inflorescence and floral induction, embryo and seedling development, root growth, fertility, trichome formation, trichome initiation, and leaf senescence (Wasternack, 2007; Browse and Howe, 2008; Gumerova et al., 2015). Jasmonates adjust plant growth and development through multifarious interconnections, constituting a complex regulatory network. In the last decades, several reports have shown that jasmonates are involved in the regulation of secondary metabolite production, like terpenoids, phytoalexins, phenylpropane, amino acid derivatives in order to activate plant defense response against biotic and abiotic stresses (Campos et al., 2014; Sharma and Laxmi, 2016; Per et al., 2018). Different researchers revealed that plant wounding and systemin enhance the endogenous level of JA and its metabolic intermediate 12-OPDA to initiate defense response machinery (Li et al., 2003; Degenhardt et al., 2010).

8.5 JASMONATE SIGNALING PATHWAY

Under multiple environmental stresses, like salt stress (Qiu et al., 2014; Zhao et al., 2014), drought, wounding, osmotic (Zhao et al., 2022) and ozone (Rao et al., 2000) stress activate the jasmonate signaling pathway in plants. Jasmonate signaling has been extensively studied in dicotyledonous plants, such as *Arabidopsis*, *Nicotiana tabacum*, and *Solanum lycopersicum* as well as in monocotyledonous plants, such as *Oryza sativa* and *Hordeum vulgare*. The jasmonate signaling cascade mechanism is well studied from *coi1* mutant analysis of *Arabidopsis*. The COI1 is identified as F-box component, which functions as E3 ubiquitin ligase of an SCF (Skp/Cullin/F-box) complex (SCF^{COI1}) responsible for jasmonate signal perception and cascade pathway. The SCF complex catalyzing the ubiquitination of specific target proteins destined for proteasomal degradation (Chini et al., 2007; Yan et al., 2007). The coronatine insensitive 1 (COI1) protein acts as the receptor for JA-Ile (bioactive form of jasmonate). In presence of hormone, COI1 recognizes JAZ (Jasmonate-ZIM domain), and facilitates the breakdown of JAZ via ubiquitin mediated-26S proteasome degradation pathway (Thines et al., 2007). JAZ acts as negative regulator in jasmonate-mediated signaling pathway (Chini et al., 2007; Thines et al., 2007). The proteasomal destruction of the JAZ protein allows transcription factors (TFs) like MYC2 to engage in downstream signaling cascade reactions. The dynamic balance between stable binding with SCF^{COI1} and its proteasomal degradation maintains the tight regulation of COI1 in plants (Figure 8.2), which is of great significance for biological function (Xie et al., 1998). The results obtained so far demonstrate that COI1 interacts with other proteins of the SCF complex, like ASK1, RBX1 and CUL1 to facilitate jasmonate-signaling responses (Xu et al., 2002; Devoto et al., 2002; Lorenzo and Solano, 2005). JAZ protein

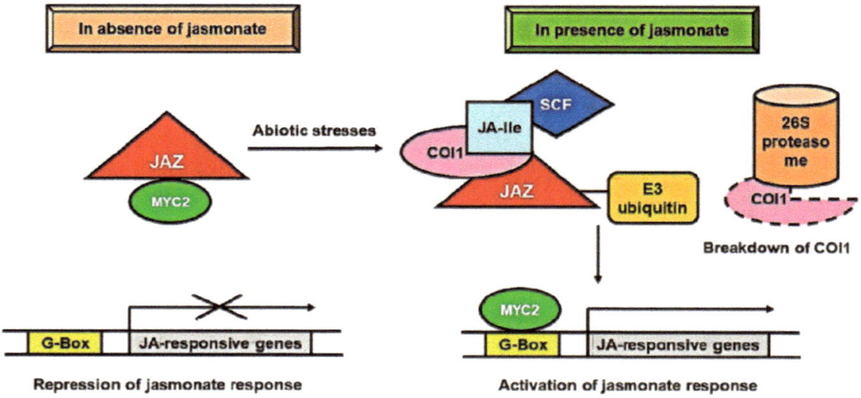

FIGURE 8.2 An integrated view of jasmonate-signaling pathway in plants.

belongs to a larger protein family TIFY protein, which includes ZIM and ZIM-like proteins that contain a zinc-finger DNA-binding domain (Shikata et al., 2003; Vanholme et al., 2007).

8.6 CROSS-TALK OF JASMONATE WITH OTHER PHYTOHORMONES

Phytohormones or PGRs control numerous physiological and developmental processes in plants and also integrate environmental and developmental cues (Kohli et al., 2013; Gómez-Cadenas et al., 2014; Jogawat, 2019; Khan et al., 2020). The cross-talk among phytohormones is intriguingly complex and their signaling pathways are interconnected with each other in a synergistic or antagonistic manner for plant sustenance under sub-optimal conditions (Verma et al., 2016). Jasmonates participate in a wide range of plant processes, which are mediated by integrated cross-talk with other major phytohormone pathways (Munemasa et al., 2011). It has no independent regulatory role but functions within a complex signal network with other phytohormones. This segment presents cross-talks on interaction outcomes of jasmonates with other phytohormones.

Studies in recent decades have revealed that jasmonates, auxin (IAA), and gibberellic acid (GA) influence each other through positive and negative regulatory interactions. The JA-IAA interactions modulate a plethora of aspects related to plant physiology. Sun et al. (2009) have documented that JA enhanced lateral root formation by inducing IAA biogenesis. JA-auxin cross-talk affects the root system development during arsenic (As) and/or cadmium (Cd) exposure in rice seedlings as revealed by Ronzan et al. (2019). MeJA enhanced the lateral root formation in the wild type and reduced the growth of seminal adventitious root under Cd and As stress, suggesting that MeJA cope up with the toxicity of Cd/As specifically in the lateral root, when the concentration of Cd/As was not so high (Ronzan et al., 2019). There is an antagonistic role between JA and GA to coordinate the developmental processes in plants under optimal and suboptimal conditions (Jang et al., 2017). JA also interacts with ABA in order to modulate tolerance to heavy metal stresses. ABA level was found to be enhanced in tomato and soybean plants by about 2.3-fold and 8.0-fold upon 24 h supplementation with JA, respectively (De Ollas et al., 2013; Wang et al., 2021). These findings pointed that JA can influence the enzymes of the ABA biosynthetic pathway as well as their respective gene expression. In a similar manner, the expression of JA biosynthetic genes and JA accumulation was also shown in *Arabidopsis* and rice (Ju et al., 2019; Wang et al., 2020). Collectively, these reports support the cross-talk between JA and ABA to facilitate plant growth. In later studies, Roselló et al. (2015) reported an induced level of both JA and ABA in Nipponbare rice root during aluminum (Al) stress, which might be required for activation of antioxidant enzymes to counteract Al-induced stress injury. JA also interacts with ethylene (ET) and auxin and their cumulative interplay during Al toxicity-induced root growth inhibition (Yang et al., 2017). In Cd-exposed *Arabidopsis* plants, the signaling pathway of ET and JA co-ordinately mediate stress-induced nitrate allocation in roots to mitigate Cd tolerance (Zhang et al., 2014).

8.7 FUNCTION OF JASMONATES IN PLANTS UNDER HEAVY METAL TOXICITY

Heavy metal contamination has become a global environmental concern and posing a severe negative impact on plant growth. Heavy metals are either completely non-essential or minutely required for the physiological functioning of plants. Plants grown on soil rich in heavy metals are known to absorb metals in quantities that are extremely toxic to plant growth and development and consequently cause overproduction of ROS and finally oxidative stress (Tchounwou et al., 2012).

Plants have shown a greater ability to adapt in exposure to heavy metal toxicity in the environment through a mechanism involving signaling molecules that can enhance stress-mediated responses. Jasmonates are one of those long-distance signaling molecules that can provide protection for plants from heavy metal/metalloid-incited phyto-toxicity by up-regulating transcript levels as well as modulating several physiological and biochemical responses. In addition, jasmonates can maintain the cellular homeostasis of plants and scavenge metal-induced excess ROS, via activating the osmolyte pool and antioxidant (enzymatic and non-enzymatic) machinery when subjected to severe stress conditions. A schematic representation of the tolerance mechanism conferred by exogenous jasmonates under heavy metal stress is represented in Figure 8.3.

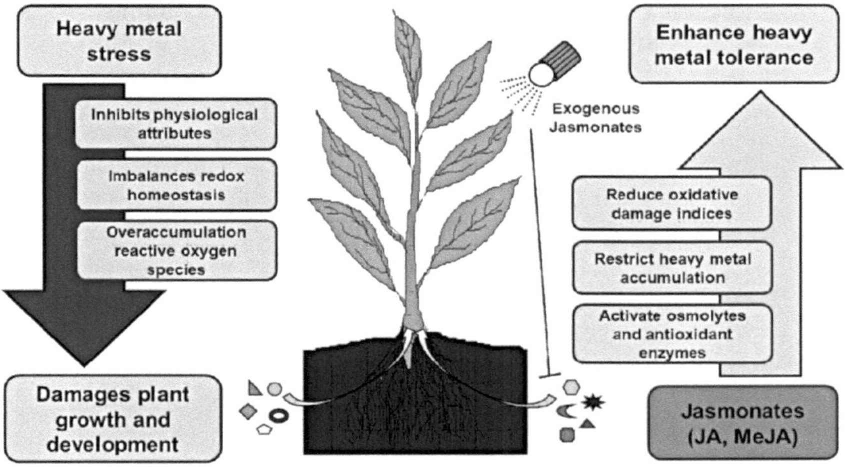

FIGURE 8.3 Generalized model illustrating the protective role of jasmonates during heavy metal stress conditions.

Some of the mechanisms involving the role of for reducing the harmful effects of heavy metals/metalloids in various plant species have been shown in Table 8.1.

1. **Cadmium:** This is one of the most hazardous elements, with a destructive impact that poses a health risk for both plant and animal systems. In Cd-exposed plants, root and shoot length was decreased, along with inhibited cell elongation, and the rate of photosynthesis was severely hampered in *Phaseolus coccineus* (Maksymiec and Krupa, 2007). It has been observed that Cd-toxicity enhanced the concentration of jasmonates in *Arabidopsis* plants (Xiang and Oliver, 1998). Tamás et al. (2009) found that the content of jasmonates was directly correlated with LOX activity under Cd-exposed environment. Treatment of MeJA was found to be effective in augmenting Cd-imparted oxidative damages in soybean plants by reducing malondialdehyde (MDA) and hydrogen peroxide (H_2O_2) content and inducing some antioxidant enzymes (Keramat et al., 2009). Previously, Singh and Shah (2014) have demonstrated that exogenous supplementation of MeJA mitigated Cd-induced oxidative injury in rice seedlings by lowering Cd-bioaccumulation in the above ground tissues that enhanced membrane integrity and ROS-homeostasis, which conferred protection of rice seedlings

towards Cd stress. This finding is similar to that obtained by Meng et al. (2009) who reported that application of 12.5 μM of JA alleviated Cd phytotoxicity in *Brassica napus* by improving chlorophyll content as well as fresh and dry mass, and lowering MDA content.

2. **Lead (Pb):** It is one of the potential pollutants with no known physiological function in plants, but is ubiquitous in nature. Bali et al. (2019) indicated that treatment of JA caused depletion in Pb uptake and boosted growth parameters in tomato plants. The combined treatment of Pb and JA displayed lowered levels of MDA, superoxide radical, and H_2O_2 in plants, with respect to only Pb-treated seedlings. Exogenous supplementation of JA played a crucial role in growth and metabolism in the aquatic plant, *Wolffia arrhiza*, which helped to avoid oxidative damage upon exposure to Pb (Piotrowska et al., 2009). Similarly, treatment of rice seedlings with MeJA reduced the Pb-induced toxicity by immobilizing Pb into the root tissues and reducing its translocation to the aerial parts, as well as reinforcing the antioxidant machinery and glyoxalase (glyoxalase I and II) response (Salavati et al., 2021). Zhu et al. (2016) provided a quantitative proteomic investigation, which demonstrated a rapid elevation of JA generation in Pb-exposed *Arabidopsis* plants, while a JA-deficient mutant and a JA-insensitive mutant displayed more pronounced root inhibition incited by Pb-phytotoxicity.
3. **Copper (Cu):** It is an essential element for plant sustainability; however, excessive use of Cu-based agrochemicals in the agricultural fields causes overaccumulation in plant tissues and consequently causes a serious issue with documented cases of phytotoxicity (Panou-Filotheou et al., 2001; Bouazizi et al., 2010). Seed priming with jasmonates showed a protective effect against Cu-toxicity by adjusting proline and enzymatic antioxidant profile [SOD, guaiacol peroxidase (POD)] and reducing stress marker (MDA) in *Cajanus cajan* seedlings (Sharma et al., 2013). Moreover, Hanaka et al. (2015) postulated that MeJA acts as a key signaling molecule in modulating physiological growth attributes in Cu-treated runner bean (*Phaseolus coccineus*) plants. Further study by Sirhindi et al. (2015) reported that seed raised from different concentrations of JA (0, 1 μM, 1 nM, and 1 pM)

was effective for eliciting Cu stress by improving photosynthetic pigments (chlorophyll and carotenoids), osmolytes (total soluble sugar, proline, and glycine betaine) and enzymatic activity of antioxidants (SOD, POD, CAT) in *Cajanus cajan* seedlings.

4. **Nickel (Ni):** It is one of the ubiquitous trace metals, but anthropogenic activities have enhanced its level in the environment. Nickel has been considered a vital micronutrient because of its requirement as a cofactor for optimum growth and development. However, excessive concentrations beyond the optimum level pose toxic effects on plant metabolism (Yusuf et al., 2011; Sreekanth et al., 2013). The stress-modulatory role of exogenous jasmonates against Ni-toxicity has been provided by Mir et al. (2018) in soybean plants. The researchers observed that JA priming improved redox homeostasis and antioxidant machinery in Ni-stressed plants by restoring ascorbate (ASC) and glutathione (GSH) content at the cellular level, which is a sign of amelioration of JA towards Ni toxicity. Moreover, JA-primed plants also hindered the uptake and translocation of Ni ion from root to shoot tissues which contributed to plant resilience, thereby indicating jasmonates as an effective stress alleviator during excessive Ni. Sirhindi et al. (2016) reported that JA facilitates soybean seedlings to combat Ni stress through modulating compatible solutes, enzymatic antioxidants [SOD, POD, ascorbate peroxidase (APX), and CAT] and their corresponding gene expression, which provide a sustainable approach to enhance crop cultivation.

5. **Arsenic:** This is a naturally occurring toxic metalloid, and its contamination is a worldwide concern due to its toxicity and carcinogenicity. Arsenic is completely non-essential and generally toxic for normal physiological functioning in plants (Samanta and Roychoudhury, 2021b). There are very few reports available to understand the response mechanism induced by jasmonates under arsenic (As) toxicity. Exogenous methyl jasmonate (MeJA) application alleviated the negative impacts of As toxicity and decreased oxidative stress by restoring chlorophyll degradation, activating antioxidant machinery, and regulating the ascorbate-glutathione (ASC-GSH) cycle in rice seedlings (Mousavi et al., 2020). Moreover, MeJA supplementation reduced bioaccumulation of As by

lowering the transcript levels of As-related transporters (*Lsi1*, *Lsi2*, and *Lsi6*), thereby improving rice growth and yield components of both cultivars (*T. hashemi* and *Fajr*). The protective role of MeJA in minimizing As-induced oxidative stress was also demonstrated by Farooq et al. (2016) in oilseed rape plants. Recently, Verma et al. (2020) demonstrated that the co-application of MeJA and As caused a marked decline in malondialdehyde (MDA) accumulation, electrolyte leakage, and bioaccumulation of total As content in rice seedlings compared to As-alone treated seedlings. The combined application also modulated the relative transcript levels of As transporters, translocation, and detoxification, revealing the probable adaptive mechanism of rice seedlings to counteract As toxicity.

6. **Chromium (Cr):** It is regarded as a serious environmental pollutant and has noxious effects on plants leading to growth and yield impairment (Shanker et al., 2005; Kumar et al., 2016). Contamination of soil and water by Cr is of recent growing concern. Recently, Kamran et al. (2021) showed the beneficial role of exogenous JA in ameliorating Cr toxicity in *Brassica parachinensis* plants. Exogenous JA mitigated the Cr-incited phytotoxic effects on photosynthetic and gas exchange parameters and enhanced the plant defense system by inducing antioxidant enzymes, glyoxalase response, and ASC-GSH pool. The JA-regulated synergistic action of antioxidant machinery and glyoxalase system restored the growth and physio-biochemical attributes in order to minimize Cr stress. In another report, supplementation of MeJA reduced the bioaccumulation of Cr, improved the photosynthetic machinery, and boosted the activity of antioxidant enzymes, which altogether diminished Cr-induced oxidative stress in pot marigold plants (Barzin et al., 2022).

7. **Aluminum:** It is the most copious metallic element in the environment. Foliar application of MeJA significantly enhanced plant tolerance to Al toxicity via stimulating lignin accumulation in the cell wall, lowering oxidative injury and escalating peroxidase activity in *Brugmansia candida* (Spollansky et al., 2000) and *Cassia tora* (Xue et al., 2008). Very recently, Esmaielzadeh et al. (2022) provided a clear insight into how exogenous MeJA

diminished the detrimental effects of Al stress on plant growth and photosynthetic apparatus by restoring mineral homeostasis and chlorophyll content. Moreover, MeJA triggered the enzymatic activity as well as glyoxalase machinery, narrowed down the toxic by-products like methylglyoxal, H_2O_2, and superoxide anion, and induced the non-protein thiol content in order to alleviate Al phytotoxicity.

8. **Boron (B):** It is an essential micronutrient, and it shows toxic symptoms when present in disproportionate concentration. Soil with toxic levels of B is widespread throughout the globe, limiting crop yield and productivity (Landi et al., 2019). Previously, Aftab et al. (2011) demonstrated that treatment of MeJA considerably improved plant growth, reduced lipid peroxidation, and induced antioxidant enzymes (SOD, CAT, POX), which collectively minimized B toxicity in *Artemisia annua* plants. Foliar application of MeJA in two Iranian grape cultivars declined MDA and proline levels, and down-regulated the expression of *VvBOR1* (encodes a B-efflux transporter) which led to more tolerance of the plant against stress situation. MeJA also enhanced the flavonoids and total phenolics content that ultimately boosted the antioxidant capacity which provided protection against B stress (Sarabandi et al., 2019). Another study by Zhao et al. (2019) showed that excess B altered the germination percentage of seeds and seedling growth in *Puccinellia tenuiflora* plants, while JA supplementation mitigated such inhibition. The study also revealed that JA restricted B uptake in plant parts, and triggered antioxidant response which was the major cause of mitigation of B-toxicity.

8.8 CONCLUDING REMARKS

A piece of explicit knowledge in metal(loid)-induced stress signaling in plants is necessary under the present scenario of climate change. However, the discovery of jasmonates has been inspected a little later than the conventional phytohormones. Various reports appraised herein advocate the control of tolerance of diverse heavy metal by this phytohormone. The major role played by the jasmonates has been elucidated in many earlier reports, uncovering that jasmonate controls diverse plant defense

TABLE 8.1 Details of Studies Exhibiting the Protective Role of Exogenous Jasmonates in Various Plant Species Against Heavy Metal/Metalloid Toxicity

Mode of Application	Plant Species	Types of Heavy Metals	Form of Jasmonate	Concentration of Jasmonate	Growth Condition	References
Exogenous supplementation	*Kandelia obovata*	Cd	MeJA	0.1, 1 and 10 $\mu mol\ L^{-1}$	Sand	Chen et al. (2014)
Exogenous supplementation	*Glycine max*	Cd	MeJA	0.01 and 0.1 mM	Perlite	Keramat et al. (2010)
Exogenous supplementation	*Glycine max*	Cd	MeJA	0.01 and 0.1 mM	Perlite	Keramat et al. (2009)
Exogenous supplementation	*Oryza sativa*	Cd	MeJA	5 µM	Sand culture	Singh and Shah (2014)
Exogenous supplementation	*Arabidopsis thaliana*	Cd	MeJA	10^{-7}, 10^{-6}, 10^{-5} M	Soil filled pots	Maksymiec et al. (2007)
Exogenous supplementation	*Solanum nigrum*	Cd	MeJA	0.01, 0.1, 10, 1000 µM	Sand filled with plastic pots	Yan et al. (2015)
Foliar spraying	*Brassica juncea*	Cd	MeJA	5, 10 and 20 µM	Earthen pots filled with acid-washed sand	Per et al. (2016)
Exogenous supplementation	*Mentha arvensis*	Cd	MeJA	1 µM	Earthen pot filled with sand and farmyard manure (6;1)	Zaid and Mohmmad (2018)
Foliar spraying	*Pisum sativum*	Cd	MeJA	1 and 10 µM	Mixture of sand and soil (3:1)	Manzoor et al. (2022)
Exogenous supplementation	*Lycopersicon esculentum*	Cd	MeJA	2.5 $\mu mol\ L^{-1}$	Hydroponic	Wei et al. (2021)
Seed pre-sowing	*Cajanus cajan*	Cd	MeJA	1 nM	Hydroponic	Kaushik et al. (2022)
Exogenous supplementation	*Arabidopsis*	Cd	MeJA	0.01 µM		Lei et al. (2019)
Exogenous supplementation	*Vicia faba*	Cd	JA	0.01 mM	Pot containing sterilized sand and vermicompost (3:1)	Ahmad et al. (2017)

TABLE 8.1 *(Continued)*

Mode of Application	Plant Species	Types of Heavy Metals	Form of Jasmonate	Concentration of Jasmonate	Growth Condition	References
Exogenous supplementation	*Avicennia marina*	Cd	JA	0.1, 1, 10, 100 $\mu mol\ L^{-1}$	Sand	Yan et al. (2015)
Seed pre-treatment	*Oryza sativa*	Cd	JA	0.2 μM	Hydroponic	Li et al. (2022)
Foliar spraying	Wheat seedlings	Cd	MeJA	0.01 and 0.1 mM	Soil and perlite	Alikhani and Abbaspour (2019)
Exogenous supplementation	Japonica rice (*Oryza sativa*)	Cd	MeJA	0.5 $\mu mol\ L^{-1}$	Hydroponic	Wu et al. (2022)
Foliar spraying	*Pisum sativum*	Cd	JA	0.5 mM	Sandy loam	Abbas et al. (2022)
Foliar spraying	*Brassica napus*	Cd	JA	25 $\mu mol\ L^{-1}$	Soil	Ali et al. (2018)
Seed priming	*Cicer arietinum*	Cd	JA	1 nM	Peat, perlite, and sand (1:1:1)	Ahmad et al. (2021)
Exogenous supplementation	*Capsicum frutescens*	Cd	MeJA	0.1, 1, 10, and 1000 $\mu mol\ L^{-1}$	Sand	Yan et al. (2013)
Exogenous supplementation	Kentucky bluegrass (*Poa pratensis*)	Cd	MeJA	75 $\mu mol\ L^{-1}$	Soil	Zhi-qiang et al. (2022)
Exogenous supplementation	Fragrant rice	Cd	MeJA	20 mM	Soil	Kanu et al. (2022)
Exogenous supplementation	*Wolffia arrhiza*	Pb	JA	0.01, 0.1, 1, 10, and 100 μM	Hutner's medium	Piotrowska et al. (2009)
Exogenous supplementation	*Oryza sativa*	Pb	MeJA	0.5 and 1 μM	Hydroponic	Salavati et al. (2021)
Seed priming	*Lycopersicon esculentum*	Pb	JA	100 nM	Hydroponic	Bali et al. (2019a)
Seed pre-soaking	*Lycopersicon esculentum*	Pb	JA	100 nM	Hydroponic	Bali et al. (2019b)
Seed pre-soaking	*Lycopersicon esculentum*	Pb	JA	0.01, 1, and 100 nM	Mixture of clay and manure (3:1)	Bali et al. (2018)

TABLE 8.1 *(Continued)*

Mode of Application	Plant Species	Types of Heavy Metals	Form of Jasmonate	Concentration of Jasmonate	Growth Condition	References
Exogenous supplementation	*Phaseolus vulgaris*	Pb	JA	100 µM	Clay and sand (2:1)	Hashem et al. (2019)
Exogenous supplementation	Cluster bean	Pb	MeJA	10, 30, and 50 µM	Hydroponic	Sharma et al. (2022)
Exogenous supplementation	*Raphanus sativus*	Pb	MeJA	0.01 and 0.1 mM	Hydroponic	Dastjerdi et al. (2015)
Foliar spraying	*Alyssum inflatum*	Ni	JA	5 and 10 µM	Perlite and sand (2:1)	Najafi Kakavand et al. (2018)
Seed priming	*Glycine max*	Ni	JA	1 nM	Peat, perlite, and sand (1:1:1)	Sirhindi et al. (2016)
Seed pre-soaking	*Glycine max*	Ni	JA	1 pM	Peat, perlite, and sand (1:1:1)	Mir et al. (2018)
Seed priming	*Glycine max*	Ni	JA	1 µM and 1 nM	Hydroponic	Sirhindi et al. (2015a)
Seed presowing	*Zea mays*	Ni	JA	10^{-6}, 10^{-8}, and 10^{-10} M	Soil	Azeem (2018)
Exogenous supplementation	*Oryza sativa*	As	MeJA	0.5 and 1 µM	Peat moss	Mousavi et al. (2020)
Exogenous supplementation	*Brassica napus*	As^{III}	MeJA	0.1 and 1 µM	Peat moss	Farooq et al. (2016)
Exogenous supplementation	*Oryza sativa*	As^{III}	MeJA	0.25 µM	Hydroponic	Verma et al. (2020)
Exogenous supplementation	*Brassica napus*	As^{III}	MeJA	0.1 and 1 µM	Peat soil	Farooq et al. (2017)
Exogenous supplementation	*Lemna valdiviana*	As^{V}	JA	50, 100, 250, and 500 µM	Clark's nutrient solution	Coelho et al. (2020)
Exogenous supplementation	*Brassica napus*	As^{III}	MeJA	0.1 and 1 µM	Peat moss	Farooq et al. (2018)
Foliar spraying	*Brassica parachinensis*	Cr	JA	5, 10, and 20 µM	Farmland topsoil and compost (3:1)	Kamran et al. (2021)
Exogenous supplementation	*Calendula officinalis*	Cr	MeJA	5 and 10 µM	Soil	Barzin et al. (2022)

TABLE 8.1 *(Continued)*

Mode of Application	Plant Species	Types of Heavy Metals	Form of Jasmonate	Concentration of Jasmonate	Growth Condition	References
Seed priming	*Cajanus cajan*	Cu	JA	1 µM, 1 nM, and 1 pM	Hydroponic	Sharma et al. (2013)
Exogenous supplementation	*Arabidopsis thaliana*	Cu	MeJA	10^{-7}, 10^{-6}, 10^{-5} M	Soil filled pots	Maksymiec et al. (2007)
Exogenous supplementation	*Phaseolus coccineus, Zea mays, Allium cepa*	Cu	MeJA	50 µM	Hydroponic	Maksymiec and Krupa (2007)
Seed pre-sowing	*Cajanus cajan*	Cu	JA	1 µM, 1 nM, and 1 pM	Hydroponic	Sirhindi et al. (2015b)
Pre-incubation	*Phaseolus coccineus*	Cu	MeJA	10 µM	Hydroponic	Hanaka et al. (2016)
Seed pre-sowing	*Cajanus cajan*	Cu	MeJA	1 nM	Hydroponic	Kaushik et al. (2022)
Exogenous supplementation	*Puccinellia tenuiflora*	B	JA	0.1, 0.5, 1, 5, 10 µM	Hydroponic	Zhao et al. (2019)
Foliar spraying	*Vitis vinifera*	B	MeJA	100 µM	Soil and sand (2:1)	Sarabandi et al. (2019)
Foliar spraying	*Artemisia annua*	B	MeJA	300 µM	Soil and farmyard manure (4:1)	Aftab et al. (2011)
Exogenous supplementation	*Oryza sativa*	Al	MeJA	0.5 and 1 µM	Hydroponic	Esmaielzadeh et al. (2022)
Foliar spraying	*Vaccinium corymbosum*	Al	MeJA	5, 10, and 50 µM	Hydroponic	Ulloa-Inostroza et al. (2017)
Exogenous supplementation	*Oryza sativa*	Se	MeJA	0.1–1.0 and 2.5–5.0 µM		Dai et al. (2021)

systems in plants. Growing evidence has revealed that exogenous supplementation as well as priming of jasmonates under heavy metal/metalloid stress can be a promising option to counteract the deleterious effects of metal-induced toxicity in plants. Although the actions of jasmonates have been well understood, several questions remain obscure. The reported investigations regarding exogenous jasmonates under heavy metal stresses are still limited. Further studies on the role of jasmonates in alleviating other heavy metal(loid) stresses should be comprehensively studied. The application of nano-biotechnology for rapid and efficient delivery of exogenous jasmonates to targeted plant cells can be utilized for the development of metal(loid) stress-resistant crop cultivars. Furthermore, future investigations can also clarify the molecular control of jasmonate signaling and cross-talk mechanism to understand the signal transduction mechanisms under heavy metal-challenged conditions to promote plant resilience, with partial improvements in plant survival and better crop yield and productivity.

KEYWORDS

- **cross-talk mechanism**
- **jasmonate**
- **phytohormone cross-talk**
- **plant growth regulators**
- **sewage disposal**
- **unsaturated fatty acid**

REFERENCES

Abbas, T., Fan, R., Hussain, S., Sattar, A., Khalid, S., Butt, M., Shahzad, U., Atif, H. M., Batool, M., Ullah, S., Li, Y., Al-Hashimi, A., & Elshikh, M. S., & Al-Yahyai, R. (2022). Protective effect of jasmonic acid and potassium against cadmium stress in peas (*Pisum sativum* L.). *Saudi Journal of Biological Sciences, 29*, 2626–2633.

Aftab, T., Khan, M. M. A., Idrees, M., Naeem, M., Moinuddin, & Hashmi, N. (2011). Methyl jasmonate counteracts boron toxicity by preventing oxidative stress and regulating antioxidant enzyme activities and artemisinin biosynthesis in *Artemisia annua* L. *Protoplasma, 248*, 601–612.

Ahmad, P., Alyemeni, M. N., Wijaya, L., Alam, P., Ahanger, M. A., & Alamri, S. A. (2017). Jasmonic acid alleviates negative impacts of cadmium stress by modifying osmolytes and antioxidants in faba bean (*Vicia faba* L.). *Archives of Agronomy and Soil Science, 63*, 1889–1899.

Ahmad, P., Raja, V., Ashraf, M., Wijaya, L., Bajguz, A., & Alyemeni, M. N. (2021). Jasmonic acid (JA) and gibberellic acid (GA) mitigated Cd-toxicity in chickpea plants through restricted Cd uptake and oxidative stress management. *Scientific Reports, 11*, 19768.

Ali, E., Hussain, N., Shamsi, I. H., Jabeen, Z., Siddiqui, M. H., & Jiang, L. (2018). Role of jasmonic acid in improving tolerance of rapeseed (*Brassica napus* L.) to Cd toxicity. *Journal of Zhejiang University Science B, 19*, 130–146.

Alikhani, O., & Abbaspour, H. (2019). Effects of methyl jasmonate and cadmium on growth traits, cadmium transport and accumulation, and allene-oxide cyclase gene expression in wheat seedlings. *Revista de Biodiversidad Neotropical, 6*, 20–29.

Avdiushko, S. A., Brown, G. C., Dahlman, D. L., & Hildebrand, D. F. (1997). Methyl jasmonate exposure induces insect resistance in cabbage and tobacco. *Environmental Entomology, 26*, 642–654.

Azeem, U. (2018). Ameliorating nickel stress by jasmonic acid treatment in *Zea mays* L. *Russian Agricultural Sciences, 44*, 209–215.

Bali, S., Jamwal, V. L., Kaur, P., Kohli, S. K., Ohri, P., Gandhi, S. G., Bhardwaj, R., Al-Huqail, A. A., Siddiqui, M. H., & Ahmad, P. (2019b). Role of P-type ATPase metal transporters and plant immunity induced by jasmonic acid against lead (Pb) toxicity in tomato. *Ecotoxicology and Environmental Safety, 174*, 283–294.

Bali, S., Jamwal, V. L., Kohli, S. K., Kaur, P., Tejpal, R., Bhalla, V., Ohri, P., Gandhi, S. G., Bhardwaj, R., Al-Huqail, A. A., Siddiqui, M. H., Ali, H. M., & Ahmad, P. (2019a). Jasmonic acid application triggers detoxification of lead (Pb) toxicity in tomato through the modifications of secondary metabolites and gene expression. *Chemosphere, 235*, 734–748.

Bali, S., Kaur, P., Kohli, S. K., Ohri, P., Thukral, A. K., Bhardwaj, R., Wijaya, L., Alyemeni, M. N., & Ahmad, P. (2018). Jasmonic acid induced changes in physio-biochemical attributes and ascorbate-glutathione pathway in *Lycopersicon esculentum* under lead stress at different growth stages. *Science of The Total Environment, 645*, 1344–1360.

Barzin, G., Safari, F., & Bishehkolaei, R. (2022). Beneficial role of methyl jasmonate on morphological, physiological and phytochemical responses of *Calendula officinalis* L. under chromium toxicity. *Physiologia Plantarum, 28*, 1453–1466.

Bouazizi, H., Jouili, H., Geitmann, A., & El Ferjani, E. (2010). Copper toxicity in expanding leaves of *Phaseolus vulgaris* L.: Antioxidant enzyme response and nutrient element uptake. *Ecotoxicology and Environmental Safety, 73*, 1304–1308.

Browse, J. (2009). Jasmonate passes muster: A receptor and targets for the defense hormone. *Annual Review of Plant Biology, 60*, 183–205.

Browse, J., & Howe, G. A. (2008). New weapons and a rapid response against insect attack. *Plant Physiology, 146*, 832–838.

Campos, M. L., Kang, J. H., & Howe, G. A. (2014). Jasmonate-triggered plant immunity. *Journal of Chemical Ecology, 40*, 657–675.

Chen, J., Yan, Z., & Li, X. (2014). Effect of methyl jasmonate on cadmium uptake and antioxidative capacity in *Kandelia obovata* seedlings under cadmium stress. *Ecotoxicology and Environmental Safety, 104*, 349–356.

Cheong, J. J., & Choi, Y. D. (2003). Methyl jasmonate as a vital substance in plants. *Trends in Plant Science, 19*, 409–413.

Chini, A., Fonseca, S., Fernández, G., Adie, B., Chico, J. M., Lorenzo, O., García-Casado, G., López-Vidriero, I., Lozano, F. M., & Ponce, M. R. (2007). The JAZ family of repressors is the missing link in jasmonate signalling. *Nature, 448*, 666–671.

Coelho, D. G., de Andrade, H. M., Marinato, C. S., Araujo, S. C., de Matos, L. P., da Silva, V. M., & de Oliveira, J. A. (2020). Exogenous jasmonic acid enhances oxidative protection of *Lemna valdiviana* subjected to arsenic. *Acta Physiologica Plantarum, 42*, 97.

Creelman, R. A., & Rao, M. V. (2002). The oxylipin pathway in *Arabidopsis*. In C. R. Somerville & E. M. Meyerowitz (Eds.), *The Arabidopsis Book*. American Society of Plant Biologists, USA.

Dai, Z., Yuan, Y., Huang, H., Hossain, M. M., Xiong, S., Cao, M., Ma, L. Q., & Tu, S. (2021). Methyl jasmonate mitigates high selenium damage of rice via altering antioxidant capacity, selenium transportation, and gene expression. *Science of The Total Environment, 756*, 143848.

Dastjerdi, Z., Afshar, A. S., & Nematpour, F. S. (2015). Effects of lead and methyl jasmonate on radish (*Raphanus sativus* L.). *Journal of Plant Processes and Functions, 4*, 59–66.

De Ollas, C., Hernando, B., Arbona, V., & Gómez-Cadenas, A. (2013). Jasmonic acid transient accumulation is needed for abscisic acid increase in citrus roots under drought stress conditions. *Physiologia Plantarum, 147*, 296–306.

Degenhardt, D. C., Refi-Hind, S., Stratmann, J. W., & Lincoln, D. E. (2010). Systemin and jasmonic acid regulate constitutive and herbivore-induced systemic volatile emissions in tomato, *Solanum lycopersicum*. *Phytochemistry, 71*, 2024–2037.

Devoto, A., Nieto-Rostro, M., Xie, D., Ellis, C., Harmston, R., Patrick, E., Davis, J., Sherratt, L., Coleman, M., & Turner, J. G. (2002). COI1 links jasmonate signaling and fertility to the SCF ubiquitin-ligase complex in *Arabidopsis*. *Plant Journal, 32*, 457–466.

Emamverdian, A., Ding, Y., & Xie, Y. (2020). The role of new members of phytohormones in plant amelioration under abiotic stress with an emphasis on heavy metals. *Polish Journal of Environmental Studies, 29*, 1009–1020.

Esmaielzadeh, S., Fallah, H., Niknejad, Y., Mahmoudi, M., & Tari, D. B. (2022). Methyl jasmonate increases aluminum tolerance in rice by augmenting the antioxidant defense system, maintaining ion homeostasis, and increasing nonprotein thiol compounds. *Environmental Science and Pollution Research, 29*, 46708–46720.

Farmer, E. E., Weber, H., & Vollenweider, S. (1998). Fatty acid signaling in *Arabidopsis*. *Planta, 206*, 167–174.

Farooq, M. A., Gill, R. A., Islam, F., Ali, B., Liu, H., Xu, J., He, S., & Zhou, W. (2016). Methyl jasmonate regulates antioxidant defense and suppresses arsenic uptake in *Brassica napus* L. *Frontiers in Plant Science, 7*, 468.

Farooq, M. A., Islam, F., Yang, C., Nawaz, A., Athar, H. R., Gill, R. A., Ali, B., Song, W., & Zhou, W. (2017). Methyl jasmonate alleviates arsenic-induced oxidative damage and modulates the ascorbate–glutathione cycle in oilseed rape roots. *Plant Growth Regulation, 84*, 135–148.

Farooq, M. A., Zhang, K., Islam, F., Wang, J., Athar, H. U. R., Nawaz, A., Ullah, Zafar Z., Xu, J., & Zhou, W. (2018). Physiological and iTRAQ-based quantitative proteomics analysis of methyl jasmonate-induced tolerance in *Brassica napus* under arsenic stress. *Proteomics*, *18*, 1700290.

Fonseca, S., Chico, J. M., & Solano, R. (2009). The jasmonate pathway: The ligand, the receptor and the core signalling module. *Current Opinion in Plant Biology*, *12*, 539–547.

Ghorbel, M., Brini, F., Sharma, A., & Landi, M. (2021). Role of jasmonic acid in plants: The molecular point of view. *Plant Cell Reports*, *40*, 1471–1494.

Ghori, N. H., Ghori, T., Hayat, M. Q., Imadi, S. R., Gul, A., Altay, V., & Ozturk, M. (2019). Heavy metal stress and responses in plants. *International Journal of Environmental Science and Technology*, *16*, 1807–1828.

Gómez-Cadenas, A., de Ollas, C., Manzi, M., & Arbona, V. (2014). Phytohormonal crosstalk under abiotic stress. In: Tran, L. S. P., & Pal, S. (Eds.) *Phytohormones: A Window to Metabolism, Signaling and Biotechnological Applications*. Springer, New York, 289–321.

Gumerova, E. A., Akulov, A. N., & Rumyantseva, N. I. (2015). Effect of methyl jasmonate on growth characteristics and accumulation of phenolic compounds in suspension culture of tartary buckwheat. *Russian Journal of Plant Physiology*, *62*, 195–203.

Hanaka, A., Wójcik, M., Dresler, S., Mroczek-Zdyrska, M., & Maksymiec, W. (2016). Does methyl jasmonate modify the oxidative stress response in *Phaseolus coccineus* treated with Cu? *Ecotoxicology and Environmental Safety*, *124*, 480–488.

Hashem, H. A., & El-Sherif, N. A. (2019). Exogenous jasmonic acid induces lead stress tolerance in kidney bean (*Phaseolus vulgaris* L.) by changing amino acid profile and stimulating antioxidant defense system. *Jordan Journal of Biological Sciences*, *12*, 345–353.

Hause, B., Stenzel, I., Miersch, O., Maucher, H., Kramell, R., Ziegler, J., & Wasternack, C. (2000). Tissue-specific oxylipin signature of tomato flowers: Allene oxide cyclase is highly expressed in distinct flower organs and vascular bundles. *Plant Journal*, *24*, 113–126.

Jang, G., Chang, S. H., Um, T. Y., Lee, S., Kim, J. K., & Choi, Y. D. (2017). Antagonistic interaction between jasmonic acid and cytokinin in xylem development. *Scientific Reports*, *7*, 10212.

Jogawat, A. (2019). Osmolytes and their role in abiotic stress tolerance in plants. In: Roychoudhury, A., & Tripathi, D. K. (Eds.) *Molecular Plant Abiotic Stress: Biology and Biotechnology*. Wiley-Blackwell, New Jersey, 91–104.

Ju, L., Jing, Y., Shi, P., Liu, J., Chen, J., Yan, J., Chu, J., Chen, K. M., & Sun, J. (2019). JAZ proteins modulate seed germination through interaction with ABI 5 in bread wheat and *Arabidopsis*. *New Phytologist*, *223*, 246–260.

Kamran, M., Wang, D., Alhaithloul, H. A. S., Alghanem, S. M., Aftab, T., Xie, K., Lu, Y., Shi, C., Sun, J., Gu, W., Xu, P., & Soliman, M. H. (2021). Jasmonic acid-mediated enhanced regulation of oxidative, glyoxalase defense system and reduced chromium uptake contributes to alleviation of chromium (VI) toxicity in choysum (*Brassica parachinensis* L.). *Ecotoxicology and Environmental Safety*, *208*, 111758.

Kanu, A. S., Ashraf, U., Mansaray, L. R., Abbas, F., Fiaz, S., Amanullah, S., Charley, C. S., & Tang, X. (2022). Exogenous methyl jasmonate application improved physio-biochemical attributes, yield, quality, and cadmium tolerance in fragrant rice. *Frontiers in Plant Science, 13*, 849477.

Kaushik, S., Sharma, P., Kaur, G., Singh, A. K., Al-Misned, F. A., Shafik, H. M., & Sirhindi, G. (2022). Seed priming with methyl jasmonate mitigates copper and cadmium toxicity by modifying biochemical attributes and antioxidants in *Cajanus cajan*. *Saudi Journal of Biological Sciences, 29*, 721–729.

Keramat, B., Kalantari, K. M., & Arvin, M. J. (2009). Effects of methyl jasmonate in regulating cadmium-induced oxidative stress in soybean plant (*Glycine max* L.). *African Journal of Microbiology Research, 3*, 240–244.

Keramat, B., Kalantari, K. M., & Arvin, M. J. (2010). Effects of methyl jasmonate treatment on alleviation of cadmium damages in soybean. *Journal of Plant Nutrition, 33*, 1016–1025.

Khan, N., Bano, A., Ali, S., & Babar, M. A. (2020). Crosstalk amongst phytohormones from planta and PGPR under biotic and abiotic stresses. *Plant Growth Regulation, 90*, 189–203.

Kohli, A., Sreenivasulu, N., Lakshmanan, P., & Kumar, P. P. (2013). The phytohormone crosstalk paradigm takes center stage in understanding how plants respond to abiotic stresses. *Plant Cell Reports, 32*, 945–957.

Kombrink, E. (2012). Chemical and genetic exploration of jasmonate biosynthesis and signaling paths. *Planta, 236*, 1351–1366.

Kumar, V., Suryakant, Sharma, P. K., Kumar, S., & Kumar, N. (2016). Effect of chromium toxicity on plants: A review. *Agriways, 4*, 107–120.

Landi, M., Margaritopoulou, T., Papadakis, I. E., & Araniti, F. (2019). Boron toxicity in higher plants: An update. *Planta, 250*, 1011–1032.

Lei, G. J., Sun, L., Sun, Y., Zhu, X. F., Li, G. X., & Zheng, S. J. (2019). Jasmonic acid alleviates cadmium toxicity in *Arabidopsis* via suppression of cadmium uptake and translocation. *Journal of Integrative Plant Biology, 62*, 218–227.

León, J., & Sánchez-Serrano, J. J. (1999). Molecular biology of jasmonic acid biosynthesis in plants. *Plant Physiology and Biochemistry, 37*, 373–380.

Li, C., Liu, G., Xu, C., Lee, G. I., Bauer, P., Ling, H. Q., Ganal, M. W., & Howe, G. A. (2003). The tomato suppressor of prosystemin-mediated responses 2 gene encodes a fatty acid desaturase required for the biosynthesis of jasmonic acid and the production of a systemic wound signal for defense gene expression. *Plant Cell, 15*, 1646–1661.

Li, Y., Zhang, S., Bao, Q., Chu, Y., Sun, H., & Huang, Y. (2022). Jasmonic acid alleviates cadmium toxicity through regulating the antioxidant response and enhancing the chelation of cadmium in rice (*Oryza sativa* L.). *Environmental Pollution, 304*, 119178.

Lorenzo, O., & Solano, R. (2005). Molecular players regulating the jasmonate signaling network. *Current Opinion in Plant Biology, 8*, 532–540.

Maksymiec, W., & Krupa, Z. (2007). Effects of methyl jasmonate and excess copper on root and leaf growth. *Biologia Plantarum, 51*, 322–326.

Maksymiec, W., Wójcik, M., & Krupa, Z. (2007). Variation in oxidative stress and photochemical activity in *Arabidopsis thaliana* leaves subjected to cadmium and excess copper in the presence or absence of jasmonate and ascorbate. *Chemosphere, 66*, 421–427.

Manara, A. (2012). Plant responses to heavy metal toxicity. In: Furini, A. (Eds.) *Plants and Heavy Metals*. Springer, Dordrecht, 27–53.

Manzoor, H., Mehwish, Bukhat, S., Rasul, S., Rehmani, M. I. A., Noreen, S., Athar, H. R., Zafar, Z. U., Skalicky, M., Soufan, W., Brestic, M., Habib-ur-Rahman, M., Ogbaga, C. C., & Sabagh, A. E. (2022). Methyl jasmonate alleviated the adverse effects of cadmium stress in pea (*Pisum sativum* L.): A nexus of photosystem II activity and dynamics of redox balance. *Frontiers in Plant Science*, *13*, 860664.

Meng, H., Hua, S., Shamsi, I. H., Jilani, G., Li, Y., & Jiang, L. (2009). Cadmium-induced stress on the seed germination and seedling growth of *Brassica napus* L., and its alleviation through exogenous plant growth regulators. *Plant Growth Regulation*, *58*, 47–59.

Mir, M. A., Sirhindi, G., Alyemeni, M. N., Alam, P., & Ahmad, P. (2018). Jasmonic acid improves growth performance of soybean under nickel toxicity by regulating nickel uptake, redox balance, and oxidative stress metabolism. *Journal of Plant Growth Regulation*, *37*, 1195–1209.

Mousavi, S. R., Niknejad, Y., Fallah, H., & Tari, D. B. (2020). Methyl jasmonate alleviates arsenic toxicity in rice. *Plant Cell Reports*, *39*, 1041–1060.

Mueller, M. J. (1997). Enzymes involved in jasmonic acid biosynthesis. *Physiologia Plantarum*, *100*, 653–663.

Munemasa, S., Mori, I. C., & Murata, Y. (2011). Methyl jasmonate signaling and signal crosstalk between methyl jasmonate and abscisic acid in guard cells. *Plant Signaling & Behavior*, *6*, 939–941.

Najafi Kakavand, S., Karimi, N., & Ghasempour, H. R. (2018). Salicylic acid and jasmonic acid restrains nickel toxicity by ameliorating antioxidant defense system in shoots of metallicolous and non-metallicolous *Alyssum inflatum* Náyr. populations. *Plant Physiology and Biochemistry*, *135*, 450–459.

Panou-Filotheou, H., Bosabalidis, A. M., & Karataglis, S. (2001). Effects of copper toxicity on leaves of oregano (*Origanum vulgare* subsp. *hirtum*). *Annals of Botany*, *88*, 207–214.

Per, T. S., Khan, M. I. R., Anjum, N. A., Masood, A., Hussain, S. J., & Khan, N. A. (2018). Jasmonates in plants under abiotic stresses: Crosstalk with other phytohormones matters. *Environmental and Experimental Botany*, *145*, 104–120.

Per, T. S., Khan, N. A., Masood, A., & Fatma, M. (2016). Methyl jasmonate alleviates cadmium-induced photosynthetic damages through increased S-assimilation and glutathione production in mustard. *Frontiers in Plant Science*, *7*, 1933.

Piotrowska, A., Bajguz, A., Godlewska-Żylkiewicz, B., Czerpak, R., & Kamińska, M. (2009). Jasmonic acid as modulator of lead toxicity in aquatic plant *Wolffia arrhiza* (Lemnaceae). *Environmental and Experimental Botany*, *66*, 507–513.

Qiu, Z., Guo, J., Zhu, A., Zhang, L., & Zhang, M. (2014). Exogenous jasmonic acid can enhance tolerance of wheat seedlings to salt stress. *Ecotoxicology and Environmental Safety*, *104*, 202–208.

Rao, M. V., Lee, H. I., Creelman, R. A., Mullet, J. E., & Davis, K. R. (2000). Jasmonic acid signaling modulates ozone-induced hypersensitive cell death. *Plant Cell*, *12*, 1633–1646.

Ronzan, M., Piacentini, D., Fattorini, L., Federica, D. R., Caboni, E., Eiche, E., Ziegler, J., House, B., Riemann, M., Betti, C., Altamura, M. M., & Falasca, G. (2019). Auxin-jasmonate crosstalk in *Oryza sativa* L. root system formation after cadmium and/or arsenic exposure. *Environmental and Experimental Botany*, *165*, 59–69.

Roselló, M., Poschenrieder, C., Gunsé, B., Barceló, J., & Llugany, M. (2015). Differential activation of genes related to aluminium tolerance in two contrasting rice cultivars. *Journal of Inorganic Biochemistry, 152*, 160–166.

Roychoudhury, A., Basu, S., & Sengupta, D. N. (2009). Comparative expression of two abscisic acid-inducible genes and proteins in the seeds of an aromatic indica rice cultivar with that of non-aromatic indica rice cultivars. *Indian Journal of Experimental Biology, 47*, 827–833.

Roychoudhury, A., Paul, S., & Basu, S. (2013). Cross-talk between abscisic acid-dependent and abscisic acid-independent pathways during abiotic stress. *Plant Cell Reports, 32*, 985–1006.

Salavati, J., Fallah, H., Niknejad, Y., & Tari, D. B. (2021). Methyl jasmonate ameliorates lead toxicity in *Oryza sativa* by modulating chlorophyll metabolism, antioxidative capacity, and metal translocation. *Physiologia Plantarum, 27*, 1089–1104.

Samanta, S., & Roychoudhury, A. (2021a). Heavy metal toxicity in plants: Physiological and molecular adaptations. In T. Aftab & K. R. Hakeem (Eds.), *Heavy Metal Toxicity in Plants* (pp. 1–10). CRC Press, Taylor & Francis.

Samanta, S., & Roychoudhury, A. (2021b). Transporters involved in arsenic uptake, translocation, and efflux in plants. In A. Roychoudhury, D. K. Tripathi, & R. Deshmukh (Eds.), *Metal and Nutrient Transporters in Abiotic Stress* (pp. 77–86). Elsevier Academic Press.

Sarabandi, M., Farokhzad, A., Mandoulakani, B. A., & Ghasemzadeh, R. (2019). Biochemical and gene expression responses of two Iranian grape cultivars to foliar application of methyl jasmonate under boron toxicity conditions. *Scientia Horticulturae, 249*, 355–363.

Shanker, A. K., Cervantes, C., Loza-Tavera, H., & Avudainayagam, S. (2005). Chromium toxicity in plants. *Environmental International, 31*, 739–753.

Sharma, M., & Laxmi, A. (2016). Jasmonates: Emerging players in controlling temperature stress tolerance. *Frontiers in Plant Science, 6*, 1129.

Sharma, P., Kaur, H., & Sirhindi, G. (2013). Effect of jasmonic acid on photosynthetic pigments and stress markers in *Cajanus cajan* (L.) Millsp. seedlings under copper stress. *American Journal of Plant Sciences, 4*, 817–823.

Sharma, Y., Kumari, N., & Sharma, V. (2022). Effect of methyl jasmonate on enhanced growth, antioxidants, and reduced Pb uptake in contrasting cluster bean cultivars. *Vegetos*. https://doi.org/10.1007/s42535-022-00491-3.

Shikata, M., Takemura, M., Yokota, A., & Kohchi, T. (2003). Arabidopsis ZIM, a plant-specific GATA factor, can function as a transcriptional activator. *Bioscience, Biotechnology, and Biochemistry, 67*, 2495–2497.

Singh, I., & Shah, K. (2014). Exogenous application of methyl jasmonate lowers the effect of cadmium-induced oxidative injury in rice seedlings. *Phytochemistry, 108*, 57–66.

Sirhindi, G., Mir, M. A., Abd-Allah, E. F., Ahmad, P., & Gucel, S. (2016). Jasmonic acid modulates the physio-biochemical attributes, antioxidant enzyme activity, and gene expression in *Glycine max* under nickel toxicity. *Frontiers in Plant Science, 7*, 591.

Sirhindi, G., Mir, M. A., Sharma, P., Gill, S. S., Kaur, H., & Mushtaq, R. (2015a). Modulatory role of jasmonic acid on photosynthetic pigments, antioxidants, and stress markers of *Glycine max* L. under nickel stress. *Physiology and Molecular Biology of Plants, 21*, 559–565.

Sirhindi, G., Sharma, P., Singh, A., Kaur, H., & Mir, M. (2015b). Alteration in photosynthetic pigments, osmolytes, and antioxidants in imparting copper stress tolerance by exogenous jasmonic acid treatment in *Cajanus cajan*. *International Journal of Plant Physiology and Biochemistry, 7*, 30–39.

Siva, G., Sivakumar, S., Prem Kumar, G., Vigneswaran, M., Vinoth, S., Muthamil Selvan, A., Parveez Ahamed, A., Manivannan, K., Rajeshkumar, R., Thajudin, N., Senthil Kumar, T., & Jayabalan, N. (2015). Optimization of elicitation conditions with jasmonic acid, characterization, and antimicrobial activity of psoralen from direct regenerated plants of *Psoralea corylifolia* L. *Biocatalysis and Agricultural Biotechnology, 4*, 624–631.

Spollansky, T. C., Pitta-Alvarez, S. I., & Giulietti, A. M. (2000). Effect of jasmonic acid and aluminum on production of tropane alkaloids in hairy root cultures of *Brugmansia candida*. *Electronic Journal of Biotechnology, 3*, 72–75.

Sreekanth, T. V. M., Nagajyothi, P. C., Lee, K. D., & Prasad, T. N. V. K. V. (2013). Occurrence, physiological responses, and toxicity of nickel in plants. *International Journal of Environmental Science and Technology, 10*, 1129–1140.

Staswick, P. E., & Tiryaki, I. (2004). The oxylipin signal jasmonic acid is activated by an enzyme that conjugates it to isoleucine in *Arabidopsis*. *Plant Cell, 16*, 2117–2127.

Sun, J., Xu, Y., Ye, S., Jiang, H., Chen, Q., Liu, F., Zhou, W., Chen, R., Li, X., Tietz, O., Wu, X., Cohen, J. D., Palme, K., & Li, C. (2009). Arabidopsis ASA1 is important for jasmonate-mediated regulation of auxin biosynthesis and transport during lateral root formation. *Plant Cell, 21*, 1495–1511.

Tamás, L., Dudíková, J., Ďurčeková, K., Halušková, L., Huttová, J., & Mistrík, I. (2009). Effect of cadmium and temperature on the lipoxygenase activity in barley root tips. *Protoplasma, 235*, 17–25.

Tchounwou, P. B., Yedjou, C. G., Patlolla, A. K., & Sutton, D. J. (2012). Heavy metal toxicity and the environment. *EXS, 101*, 133–164.

Thines, B., Katsir, L., Melotto, M., Niu, Y., Mandaokar, A., Liu, G., Nomura, K., He, S. Y., Howe, G. A., & Browse, J. (2007). JAZ repressor proteins are targets of the SCFCOI1 complex during jasmonate signaling. *Nature, 448*, 661–665.

Ulloa-Inostroza, E. M., Alberdi, M., Meriño-Gergichevich, C., & Reyes-Díaz, M. (2017). Low doses of exogenous methyl jasmonate applied simultaneously with toxic aluminum improve the antioxidant performance of *Vaccinium corymbosum*. *Plant and Soil, 412*, 81–96.

Vanholme, B., Grunewald, W., Bateman, A., Kohchi, T., & Gheysen, G. (2007). The TIFY family previously known as ZIM. *Trends in Plant Science, 12*, 239–244.

Verma, G., Srivastava, D., Narayan, S., Shirke, P. A., & Chakrabarty, D. (2020). Exogenous application of methyl jasmonate alleviates arsenic toxicity by modulating its uptake and translocation in rice (*Oryza sativa* L.). *Ecotoxicology and Environmental Safety, 201*, 110735.

Verma, V., Ravindran, P., & Kumar, P. P. (2016). Plant hormone-mediated regulation of stress responses. *BMC Plant Biology, 16*, 86.

Vijayan, P., Shockey, J., Levesque, C. A., Cook, R. J., & Browse, J. (1998). A role for jasmonate in pathogen defense of *Arabidopsis*. *Proceedings of the National Academy of Sciences, 95*, 7209–7214.

Waldemar, M., Małgorzata, W., & Zbigniew, K. (2007). Variation in oxidative stress and photochemical activity in *Arabidopsis thaliana* leaves subjected to cadmium and excess copper in the presence or absence of jasmonate and ascorbate. *Chemosphere*, *66*, 421–427.

Wang, J., Song, L., Gong, X., Xu, J., & Li, M. (2020). Functions of jasmonic acid in plant regulation and response to abiotic stress. *International Journal of Molecular Sciences*, *21*, 1446.

Wang, X., Li, Q., Xie, J., Huang, M., Cai, J., Zhou, Q., Dai, T., & Jiang, D. (2021). Abscisic acid and jasmonic acid are involved in drought priming-induced tolerance to drought in wheat. *Crop Journal*, *9*, 120–132.

Wang, Y., Hou, Y., Qiu, J., Wang, H., Wang, S., Tang, L., Tong, X., & Zhang, J. (2020). Abscisic acid promotes jasmonic acid biosynthesis via a 'SAPK10-bZIP72-AOC' pathway to synergistically inhibit seed germination in rice (*Oryza sativa*). *New Phytologist*, *228*, 1336–1353.

Wasternack, C. (2007). Jasmonates: An update on biosynthesis, signal transduction and action in plant stress response, growth and development. *Annals of Botany*, *100*, 681–697.

Wasternack, C., & Hause, B. (2002). Jasmonates and octadecanoids: Signals in plant stress responses and development. *Progress in Nucleic Acid Research and Molecular Biology*, *72*, 165–221.

Wasternack, C., & Hause, B. (2013). Jasmonates: Biosynthesis, perception, signal transduction and action in plant stress response, growth and development. *Annals of Botany*, *111*, 1021–1068.

Wasternack, C., & Strnad, M. (2016). Jasmonate signaling in plant stress responses and development–Active and inactive compounds. *New Biotechnology*, *33*, 604–613.

Wei, T., Li, X., Yashir, N., Li, H., Sun, Y., Hua, L., Ren, X., & Guo, J. (2021). Effect of exogenous silicon and methyl jasmonate on the alleviation of cadmium-induced phytotoxicity in tomato plants. *Environmental Science and Pollution Research*, *28*, 51854–51864.

Wu, B., Zeng, Z., Wu, X., Li, Y., Wang, F., Yang, J., & Li, X. (2022). Jasmonic acid negatively regulates root growth in Japonica rice (*Oryza sativa* L.) under cadmium treatment. *Plant Growth Regulation*, *98*, 651–667.

Xiang, C., & Oliver, D. J. (1998). Glutathione metabolic genes coordinately respond to heavy metals and jasmonic acid in *Arabidopsis*. *Plant Cell*, *10*, 1539–1550.

Xie, D. X., Feys, B. F., James, S., Nieto-Rostro, M., & Turner, J. G. (1998). COI1: An *Arabidopsis* gene required for jasmonate-regulated defense and fertility. *Science*, *280*, 1091–1094.

Xu, L., Liu, F., Lechner, E., Genschik, P., Crosby, W. L., Ma, H., Peng, W., Huang, D., & Xie, D. (2002). The SCFCOI1 ubiquitin-ligase complexes are required for jasmonate response in *Arabidopsis*. *Plant Cell*, *14*, 1919–1935.

Xue, Y. J., Tao, L., & Yang, Z. H. (2008). Aluminum-induced cell wall peroxidase activity and lignin synthesis are differentially regulated by jasmonate and nitric oxide. *Journal of Agricultural and Food Chemistry*, *56*, 9676–9684.

Yan, Y., Stolz, S., Chetelat, A., Reymond, P., Pagni, M., Dubugnon, L., & Farmer, E. E. (2007). A downstream mediator in the growth repression limb of the jasmonate pathway. *Plant Cell*, *19*, 2470–2483.

Yan, Z., Chen, J., & Li, X. (2013). Methyl jasmonate as modulator of Cd toxicity in *Capsicum frutescens* var. *fasciculatum* seedlings. *Ecotoxicology and Environmental Safety*, *98*, 203–209.

Yan, Z., Li, X., Chen, J., & Tam, N. F. Y. (2015). Combined toxicity of cadmium and copper in *Avicennia marina* seedlings and the regulation of exogenous jasmonic acid. *Ecotoxicology and Environmental Safety*, *113*, 124–132.

Yan, Z., Zhang, W., Chen, J., & Li, X. (2015). Methyl jasmonate alleviates cadmium toxicity in *Solanum nigrum* by regulating metal uptake and antioxidative capacity. *Biologia Plantarum*, *59*, 373–381.

Yang, Z. B., He, C., Ma, Y., Herde, M., & Ding, Z. (2017). Jasmonic acid enhances Al-induced root growth inhibition. *Plant Physiology*, *173*, 1420–1433.

Yu, X., Zhang, W., Zhang, Y., Zhang, X., Lang, D., & Zhang, X. (2018). The roles of methyl jasmonate to stress in plants. *Functional Plant Biology*, *46*, 197–212.

Yusuf, M., Fariduddin, Q., Hayat, S., & Ahmad, A. (2011). Nickel: An overview of uptake, essentiality and toxicity in plants. *Bulletin of Environmental Contamination and Toxicology*, *86*, 1–17.

Zaid, A., & Mohammad, F. (2018). Methyl jasmonate and nitrogen interact to alleviate cadmium stress in *Mentha arvensis* by regulating physio-biochemical damages and ROS detoxification. *Journal of Plant Growth Regulation*, *37*, 1331–1348.

Zavala, J. A., & Baldwin, I. T. (2006). Jasmonic acid signalling and herbivore resistance traits constrain regrowth after herbivore attack in *Nicotiana attenuata*. *Plant, Cell & Environment*, *29*, 1751–1760.

Zhang, G. B., Yi, H. Y., & Gong, J. M. (2014). The *Arabidopsis* ethylene/jasmonic acid-NRT signaling module coordinates nitrate reallocation and the trade-off between growth and environmental adaptation. *Plant Cell*, *26*, 3984–3998.

Zhao, Q., Sun, Q., Dong, P., Ma, C., Sun, H., & Liu, C. (2019). Jasmonic acid alleviates boron toxicity in *Puccinellia tenuiflora*, a promising species for boron phytoremediation. *Plant and Soil*, *445*, 397–407.

Zhao, W., Huang, H., Wang, J., Wang, X., Xu, B., Yao, X., Sun, L., Yang, R., Wang, J., Sun, A., & Wang, S. (2022). Jasmonic acid enhances osmotic stress responses by MYC2-mediated inhibition of protein phosphatase 2C1 and response regulators 26 transcription factor in tomato. *Plant Journal*, *113*, 546–561.

Zhao, Y., Dong, W., Zhang, N., Ai, X., Wang, M., Huang, Z., Xiao, L., & Xia, G. (2014). A wheat allene oxide cyclase gene enhances salinity tolerance via jasmonate signaling. *Plant Physiology*, *164*, 1068–1076.

Zhi-qiang, M. A., Yi, S., & Pei-fang, Z. (2022). Effects of exogenous methyl jasmonate on seed germination and seedling growth of Kentucky bluegrass under cadmium pollution. *Acta Agron Sin*, *30*, 1475–1484.

Zhu, F. Y., Chan, W. L., Chen, M. X., Kong, R. P. W., Cai, C., Wang, Q., Zhang, J. H., & Lo, C. (2016). SWATH-MS quantitative proteomic investigation reveals a role of jasmonic acid during lead response in *Arabidopsis*. *Journal of Proteome Research*, *15*, 3528–3539.

Index

T